AF684598

ELECTROCHEMISTRY AND ELECTROCHEMICAL ENGINEERING

Edited by
Levitah Castil Mapatac

Delve Publishing

Electrochemistry and Electrochemical Engineering

Edited by Levitah Castil Mapatac

© 2017

Edition 2018

Delve Publishing

Published by

Delve Publishing LLC

708 3rd Avenue, 6th Floor

New York

NY 10801

United States of America

www.delvepublishing.com

ISBN: 978-1-68095-723-5 (Hardback)

Printed in United States of America

Library of Congress Control Number: 2016920403

This book contains information obtained from highly regarded resources. Some content has been published with permission under Creative Commons License or similar. A Wide variety of references are listed. Reasonable efforts have been made to publish reliable data and views articulated in the chapters are those of the individual contributors, and not necessarily those of the editors or publishers. Editors or publishers are not responsible for the accuracy of the information in the published chapters or consequences of their use. The publisher believes no responsibility for any damage or grievance to the persons or property arising out of the use of any materials, instructions, methods or thoughts in the book. The editors and the publisher have attempted to trace the copyright holders of all material reproduced in this publication and apologize to copyright holders if permission has not been obtained. If any copyright holder has not been acknowledged, please write to us so we may rectify.

Notice: Registered trademark of products or corporate names are used only for explanation and identification without intent of infringement. Although care has been taken to check accuracy of formulas and procedures, the detailed methods should be tested further on a small scale before being adopted commercially.

Delve Publishing publishes wide variety of books and eBooks. For more information about Delve Publishing and its products, visit our website at www.delvepublishing.com

About the Editor

Levitah Castil Mapatac

Levitah C. Mapatac is an Associate Professor, researcher, writer and a PhD holder from the University in Science and Technology of Southern Philippines. Her interest varies from different chemistry concepts and applications and currently doing research in traditional medicines, water waste pollution and environmental chemistry in the aspect of contamination pathways, profilling and rehabilatation of mining waste.

Table of Contents

1 ELECTROCHEMISTRY PRINCIPLES AND APPLICATIONS

2 ELECTRODE POTENTIAL

3 ELECTRODE POTENTIAL

4 ELECTROLYTIC CONDUCTIVITY

5 ELECTROANALYTICAL METHODS

6 ELECTRON TRANSFER MECHANISM

7 ELECTROCHEMICAL REACTION MECHANISM

8 ELECTROCHEMICAL CELL

9 ELECTROHYDROGENESIS

10 BIOELECTROCHEMISTRY

List of Abbreviations

TMS	Causing Transcranial Magnetic Stimulation
CNC	Computer Numerical Controlled
CEDI	Continuous Electrodeionization
CV	Cyclic Voltammetry
DME	Dropping Mercury Electrode
EDM	Electrical Discharge Machining
ECM	Electrochemical Machining
ECT	Electroconvulsive Therapy
ET	Electron Transfer
EET	extracellular Electron Transfer
ELF	Extremely Low Frequency
GEC	Graphite-Epoxy Composite
HMDE	Hanging mercury drop Electrode
HPLC	High Performance Liquid Chromotography
IS	Inner Sphere
ISE	International Society of Electrochemistry
MEC	Microbial Electrolysis Cells
MFC	Microbial Fuel Cell
PMT	Photomultiplier Tubes
PAGE	Polyacrylamide Gel Electrophoresis
POIPCs	Protic Organic Ionic Plastic Crystals
PEM	Proton Exchange Membrane

RDE	Rotating Disk Electrode
RRDE	Rotating Ring-Disk Electrode
SB	Schiff Base
SFFF	Sedimentation Field Flow Fractionation
SWV	Squarewave Voltammetry
SHE	Standard Hydrogen Electrode
TENS	Transcutaneous Electrical Nerve Stimulation
UME	Ultramicroelectrode
WEDM	Wire Electrical Discharge Machining

Preface

The science of electrochemistry was born at the end of the eighteenth century following intense interest in the production of electricity by animals such as the torpedo and the electric eel. The key phenomena were observed just at the turn of the century. Volta invented his pile in which chemical species were consumed with the production of electricity and a few months afterwards Nicholson and Carlisle used this electricity to decompose water into hydrogen and oxygen. Electrochemistry is the branch of physical chemistry that studies the relationship between electricity, as a measurable and quantitative phenomenon, and identifiable chemical change, with either electricity considered an outcome of a particular chemical change or vice versa. These reactions involve electric charges moving between electrodes and an electrolyte. Thus electrochemistry deals with the interaction between electrical energy and chemical change. Electrochemistry continues to find new applications in wider fields as well as providing scope for greater understanding of its fundamentals. Electrochemical Engineering sounds very much like chemical engineering, but the chemists, electro chemists, material scientists and whoever else comes into touch with technical electrochemical systems very soon gets the feeling, that chemical engineering wisdom will not get them very far in enhancing their understanding and helping them to solve their problems with technical electro chemical devices. Indeed not only the appearance of but also the physics and physical chemistry in electrochemical reactors - electrolyzers, batteries or fuel cells and others - are quite different from that of normal chemical reactors.

Content Coverage

Chapter One presents the introduction of electrochemistry that includes the study of chemical changes caused by the passage of an electric current across a medium, as well as the production of electric energy by chemical reactions. Electrochemistry also embraces the study of electrolyte solutions and the chemical equilibria that occur in them.

Chapter Two focuses on electrode potential, which is the electric potential on an electrode component. In a cell, there will be an electrode potential for the cathode, and an electrode potential for the anode.

Chapter Three focuses primarily on electrokinetic phenomena that occur in heterogeneous fluids, or in porous bodies filled with fluid, or in a fast flow over a flat surface. The term heterogeneous here means a fluid containing particles.

Chapter Four highlights about electrolytic conductivity measurements used routinely in many industrial and environmental applications as a fast, inexpensive and reliable way of measuring the ionic content in a solution.

Chapter Five focuses on electroanalytical methods, a class of techniques in analytical chemistry which study an analyte by measuring the potential (volts) and/or current (amperes) in an electrochemical cell containing the analyte.

Chapter Six emphasizes on the mechanism of electron transfer (ET) that occurs when an electron relocates from an atom or molecule to another such chemical entity. ET is a mechanistic description of a redox reaction, wherein the oxidation state of reactant and product changes.

Chapter Seven focuses on 'electrochemical mechanisms important to all redox chemistry including corrosion, redox active photochemistry including photosynthesis, other biological systems often involving electron transport chains and other forms of homogeneous and heterogeneous electron transfer.

Chapter Eight presents an approach on electrochemical cell capable of either generating electrical energy from chemical reactions or facilitating chemical reactions through the introduction of electrical energy.

Chapter Nine explores the process of electrohydrogenesis for generating hydrogen gas from organic matter being decomposed by bacteria.

Chapter Ten is about bioelectromagnetics that deals with the study of the interaction between electromagnetic fields and biological entities.

CHAPTER 1

Electrochemistry Principles and Applications

Chapter Contents

- **Introduction**
- **History of Electrochemistry**
- **Principless**
- **Battery**
- **Corrosion**
- **Applications**

Electrochemistry deals with the links between chemical reactions and electricity. This includes the study of chemical changes caused by the passage of an electric current across a medium, as well as the production of electric energy by chemical reactions. Electrochemistry also embraces the study of electrolyte solutions and the chemical equilibria that occur in them.

Many chemical reactions require the input of energy. Such reactions can be carried out at the surfaces of electrodes in cells connected to external power supplies. These reactions provide information about the nature and properties of the chemical species contained in the cells, and can also be used to synthesize new chemicals. The production of chlorine and aluminum and the electroplating and electrowinning of metals are examples of industrial electrochemical processes. Electrochemical cells that produce electric energy from chemical energy are the basis of primary and secondary (storage) batteries and fuel cells. Other electrical phenomena of interest in chemical systems include the behavior of ionic solutions and the conduction of current through these solutions, the separation of ions by an electric field (electrophoresis), the corrosion and passivation of metals, electrical effects in biological systems (bioelectrochemistry), and the effect of light on electrochemical cells (photoelectrochemistry).

INTRODUCTION

Electrochemistry is the branch of physical chemistry that studies the relationship between electricity, as a measurable and quantitative phenomenon, and identifiable chemical change, with either electricity considered an outcome of a particular chemical change or vice versa. These reactions involve electric charges moving between electrodes and an electrolyte (or ionic species in a solution). Thus electrochemistry deals with the interaction between electrical energy and chemical change.

When a chemical reaction is caused by an externally supplied current, as in electrolysis, or if an electric current is produced by a spontaneous chemical reaction as in a battery, it is called an *electrochemical* reaction. Chemical reactions where electrons are transferred directly between molecules and/or atoms are called oxidation-reduction or (redox) reactions. In general, electrochemistry describes the overall reactions when individual redox reactions are separate but connected by an external electric circuit and an intervening electrolyte.

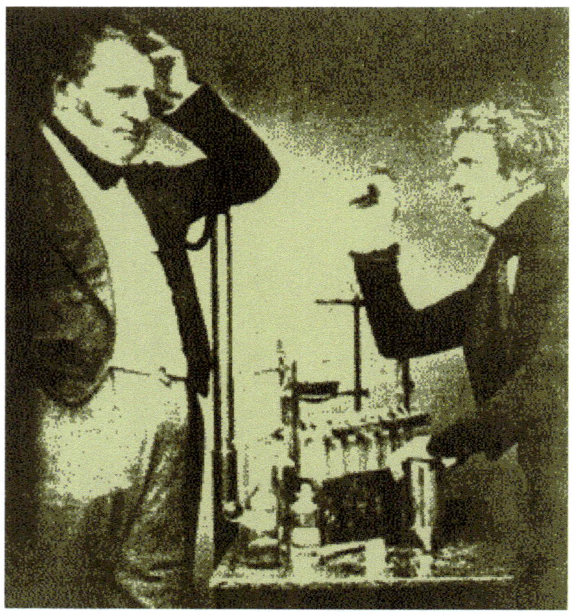

English chemist John Daniell (left) and physicist Michael Faraday (right), both credited as founders of electrochemistry today.

HISTORY OF ELECTROCHEMISTRY

The 16th century marked the beginning of scientific understanding of electricity and magnetism that culminated with the production of electric power and the industrial revolution in the late 19th century

In the 1550s, English scientist William Gilbert spent 17 years experimenting with magnetism and, to a lesser extent, electricity. For his work on magnets, Gilbert became known as "The Father of Magnetism." His book *De Magnete* quickly became the standard work throughout Europe on electrical and magnetic clear distinction between magnetism and what was then called the "amber effect" (static electricity).

German physicist Otto von Guericke beside his electrical generator while conducting experiment.

In 1663, German physicist Otto von Guericke created the first electrostatic generator, which produced static electricity by applying friction. The generator was made of a large sulfur ball inside a glass globe, mounted on a shaft. The ball was rotated by means of a crank and a static electricspark was produced when a pad was rubbed against the ball as it rotated. The globe could be removed and used as an electrical source for experiments with electricity. Von Guericke used his generator to show that like charges repelled each other.

The 18th Century and Birth of Electrochemistry

In 1709, Francis Hauksbee at the Royal Society in London discovered that by putting a small amount of mercury in the glass of Von Guericke's generator and evacuating the air from it, it would glow whenever the ball built up a charge and his hand was touching the globe. He had created the first gas-discharge lamp.

Between 1729 and 1736, two English scientists, Stephen Gray and Jean Desaguliers, performed a series of experiments which showed that a cork or other object as far away as 800 or 900 feet (245–275 m) could be electrified by connecting it via a charged glass tube to materials such as metal wires or hempen string. They found that other materials, such as silk, would not convey the effect.

By the mid-18th century, French chemist Charles François de Cisternay Du Fay had discovered two forms of static electricity, and that like charges repel each other while unlike charges attract. Du Fay announced that electricity consisted of two fluids: *vitreous* (from the Latin for "glass"), or positive, electricity; and *resinous*, or negative, electricity. This was the "two-fluid theory" of electricity, which was opposed by Benjamin Franklin's "one-fluid theory" later in the century.

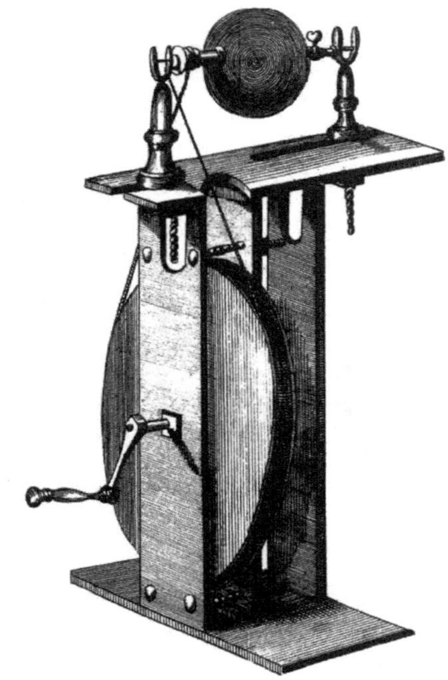

Francis Hauksbee's gas-discharge lamp.

In 1745, Jean-Antoine Nollet developed a theory of electrical attraction and repulsion that supposed the existence of a continuous flow of electrical matter between charged bodies. Nollet's theory at first gained wide acceptance, but met resistance in 1752 with the translation of Franklin's *Experiments and Observations on Electricity* into French. Franklin and Nollet debated the nature of electricity, with Franklin supporting action at a distance and two qualitatively opposing types of electricity, and Nollet advocating mechanical action and a single type of electrical fluid. Franklin's argument eventually won and Nollet's theory was abandoned.

In 1748, Nollet invented one of the first electrometers, the electroscope, which showed electric charge using electrostatic attraction and repulsion. Nollet is reputed to be the first to apply the name "Leyden jar" to the first device for storing electricity. Nollet's invention was replaced by Horace-Bénédict de Saussure's electrometer in 1766.

By the 1740s, William Watson had conducted several experiments to determine the speed of electricity. The general belief at the time was that electricity was faster than sound, but no accurate test been devised to measure the velocity of a current. Watson, in the fields north of London, laid out a line of wire supported by dry sticks and silk which ran for 12,276 feet (3.7 km). Even at this length, the velocity of electricity seemed instantaneous. Resistance in the wire was also noticed but apparently not fully understood, as Watson related that "we observed again, that although the electrical compositions were very severe to those who held the wires, the report of the Explosion at the prime Conductor was little, in comparison of that which is heard when the Circuit is short." Watson eventually decided not to pursue his electrical experiments, concentrating instead upon his medical career.

By the 1750s, as the study of electricity became popular, efficient ways of producing electricity were sought. The generator developed by Jesse Ramsden was among the first electrostatic generators invented. Electricity produced by such generators was used to treat paralysis, muscle spasms, and to control heart rates. Other medical uses of electricity included filling the body with electricity, drawing sparks from the body, and applying sparks from the generator to the body.

Charles-Augustin de Coulomb developed the law of electrostatic attraction in 1781 as an outgrowth of his attempt to investigate the law of electrical repulsions as stated by Joseph Priestley in England. To this end, he invented a sensitive apparatus to measure the electrical forces involved in Priestley's law. He also established the inverse square law of attraction and repulsion magnetic poles, which became the basis for the mathematical theory of magnetic forces developed by Siméon Denis Poisson. Coulomb wrote seven important works on electricity and magnetism which he submitted to the Académie des Sciences between 1785 and 1791, in which he reported having developed a theory of attraction and repulsion between charged bodies, and went on to search for perfect conductors and dielectrics. He suggested that there was no perfect dielectric, proposing that every substance has a limit, above which it will conduct electricity. The SI unit of charge is called a coulomb in his honor.

In 1789, Franz Aepinus developed a device with the properties of a "condenser" (now known as a capacitor.) The Aepinus condenser was the first capacitor developed after the Leyden jar, and was used to demonstrate conduction and induction. The device was constructed so that the space between two plates could be adjusted, and the glass dielectric separating the two plates could be removed or replaced with other materials.

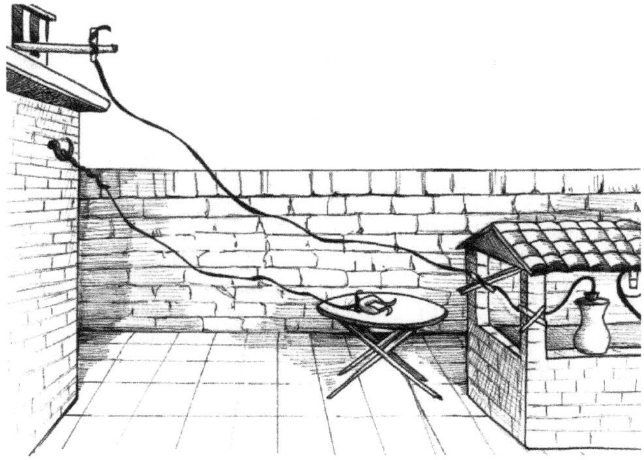

Late 1780s diagram of Galvani's experiment on frog legs.

Despite the gain in knowledge of electrical properties and the building of generators, it wasn't until the late 18th century that Italian physician and anatomist Luigi Galvani marked the birth of electrochemistry by establishing a bridge between muscular contractions and electricity with his 1791 essay *De Viribus Electricitatis in Motu Musculari Commentarius* (Commentary on the Effect of Electricity on Muscular Motion), where he proposed a "nerveo-electrical substance" in life forms.

In his essay, Galvani concluded that animal tissue contained a before-unknown innate, vital force, which he termed "animal electricity," which activated muscle when placed between two metal probes. He believed that this was evidence of a new form of electricity, separate from the "natural" form that is produced by lightning and the "artificial" form that is produced by friction (static electricity). He considered the brain to be the most important organ for the secretion of this "electric fluid" and that the nerves conducted the fluid to the muscles. He believed the tissues acted similarly to the outer and inner surfaces of Leyden jars. The flow of this electric fluid provided a stimulus to the muscle fibers.

Italian physicist Alessandro Volta showing his *"battery"* to French emperor Napoleon Bonaparte in the early 19th century.

Galvani's scientific colleagues generally accepted his views, but Alessandro Volta, the outstanding professor of physics at the University of Pavia, was not convinced by the analogy between muscles and Leyden jars. Deciding that the frogs' legs used in Galvani's experiments served only as an electroscope, he held that the contact of dissimilar metals was the true source of stimulation. He referred to the electricity so generated as "metallic electricity" and decided that the muscle, by contracting when touched by metal, resembled the action of an electroscope. Furthermore, Volta claimed that if two dissimilar metals in contact with each other also touched a muscle, agitation would also occur and increase with the dissimilarity of the metals. Galvani refuted this by obtaining muscular action using two pieces of similar metal. Volta's name was later used for the unit of electrical potential, the volt.

Rise of Electrochemistry as Branch of Chemistry

In 1800, English chemists William Nicholson and Johann Wilhelm Ritter succeeded in separating water into hydrogen and oxygen by electrolysis. Soon thereafter, Ritter discovered the process of electroplating. He also observed that the amount of metal deposited and the amount of oxygen produced during an electrolytic process depended on the distance between the electrodes. By 1801 Ritter had observed thermoelectric currents, which anticipated the discovery of thermoelectricity by Thomas Johann Seebeck.

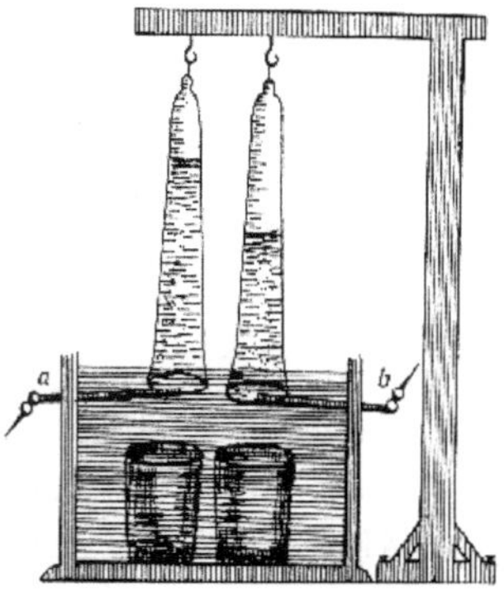

Scheme of Ritter's apparatus to separate water into hydrogen and oxygen by electrolysis.

In 1802, William Cruickshank designed the first electric battery capable of mass production. Like Volta, Cruickshank arranged square copper plates, which he soldered at their ends, together with plates of zinc of equal size. These plates were placed into a long rectangular wooden box which was sealed with cement. Grooves inside the box held the metal plates in position. The box was then filled with an electrolyte of brine, or watered down acid. This flooded design had the advantage of not drying out with use and provided more energy than Volta's arrangement, which used brine-soaked papers between the plates.

In the quest for a better production of platinum metals, two scientists, William Hyde Wollaston and Smithson Tennant, worked together to design an efficient electrochemical technique to refine or purify platinum. Tennant ended up discovering the elements iridium and osmium. Wollaston's effort, in turn, led him to the discovery of the metals palladium in 1803 and rhodium in 1804.

Wollaston made improvements to the galvanic battery (named after Galvani) in the 1810s. In Wollaston's battery, the wooden box was replaced with an earthenware vessel, and a copper plate was bent into a U-shape, with a single plate of zinc placed in the center of the bent copper. The zinc plate was prevented from making contact with the copper by dowels (pieces) of cork or wood. In his single cell design, the U-shaped copper plate was welded to a horizontal handle for lifting the copper and zinc plates out of the electrolyte when the battery was not in use.

In 1809, Samuel Thomas von Soemmering developed the first telegraph. He used a device with 26 wires (1 wire for each letter of the German alphabet) terminating in a container of acid. At the sending station, a key, which completed a circuit with a battery, was connected as required to each of the line wires. The passage of current caused the acid to decompose chemically, and the message was read by observing at which of the terminals the bubbles of gas appeared. This is how he was able to send messages, one letter at a time.

Humphry Davy's work with electrolysis led to conclusion that the production of electricity in simple electrolytic cells resulted from chemical reactions between the electrolyte and the metals, and occurred between substances of opposite charge. He reasoned that the interactions of electric currents with chemicals offered the most likely means of decomposing all substances to their basic elements. These views were explained in 1806 in his lecture *On Some Chemical Agencies of Electricity*, for which he received the Napoleon Prize from the Institut de France in 1807 (despite the fact that England and France were at war at the time). This work led directly to the isolation of sodium and potassium from their common compounds and of the alkaline earth metals from theirs in 1808.

Hans Christian Ørsted's discovery of the magnetic effect of electric currents in 1820 was immediately recognized as an important advance, although he left further work on electromagnetism to others. André-Marie Ampère quickly repeated Ørsted's experiment, and formulated them mathematically (which became Ampère›s law). Ørsted also discovered that not only is a magnetic needle deflected by the electric current, but that the live electric wire is also deflected in a magnetic field, thus laying the foundation for the construction of an electric motor. Ørsted's discovery of piperine, one of the pungent components of pepper, was an important contribution to chemistry, as was his preparation of aluminium in 1825.

During the 1820s, Robert Hare developed the Deflagrator, a form of voltaic battery having large plates used for producing rapid and powerful combustion. A modified form of this apparatus was employed in 1823 in volatilising and fusing carbon. It was with these batteries that the first use of voltaic electricity for blasting under water was made in 1831.

In 1821, the Estonian-German physicist, Thomas Johann Seebeck, demonstrated the electrical potential in the juncture points of two dissimilar metals when there is a temperature difference between the joints. He joined a copper wire with a bismuth wire to form a loop or circuit. Two junctions were formed by connecting the ends of the wires to each other. He then accidentally discovered that if he heated one junction to a high temperature, and the other junction remained at room temperature, a magnetic field was observed around the circuit.

He did not recognize that an electric current was being generated when heat was applied to a bi-metal junction. He used the term "thermomagnetic currents" or "thermomagnetism" to express his discovery. Over the following two years, he reported on his continuing observations to the Prussian Academy of Sciences, where he described his observation as "the magnetic polarization of metals and ores produced by a temperature difference." This Seebeck effect became the basis of the thermocouple, which is still considered the most accurate measurement of temperature today. The converse Peltier effect was seen over a decade later when a current was run through a circuit with two dissimilar metals, resulting in a temperature difference between the metals.

In 1827 German scientist Georg Ohm expressed his law in his famous book *Die galvanische Kette, mathematisch bearbeitet* (The Galvanic Circuit Investigated Mathematically) in which he gave his complete theory of electricity.

In 1829 Antoine-César Becquerel developed the "constant current" cell, forerunner of the well-known Daniell cell. When this acid-alkali cell was monitored by a galvanometer, current was found to be constant for an hour, the first instance of "constant current". He applied the results of his study of thermoelectricity to the construction of an electric thermometer, and measured the temperatures of the interior of animals, of the soil at different depths, and of the atmosphere at different heights. He helped validate Faraday's laws and conducted extensive investigations on the electroplating of metals with applications for metal finishing and metallurgy. Solar cell technology dates to 1839 when Becquerel observed that shining light on an electrode submerged in a conductive solution would create an electric current.

Michael Faraday began, in 1832, what promised to be a rather tedious attempt to prove that all electricities had precisely the same properties and caused precisely the same effects. The key effect was electrochemical decomposition. Voltaic and electromagnetic electricity posed no problems, but static electricity did. As Faraday delved deeper into the problem, he made two startling discoveries. First, electrical force did not, as had long been supposed, act at a distance upon molecules to cause them to dissociate. It was the passage of electricity through a conducting liquid medium that caused the molecules to dissociate, even when the electricity merely discharged into the air and did not pass through a "pole" or "center of action" in a voltaic cell. Second, the amount of the decomposition was found to be related directly to the amount of electricity passing through the solution.

These findings led Faraday to a new theory of electrochemistry. The electric force, he argued, threw the molecules of a solution into a state of tension. When the force was strong enough to distort the forces that held the molecules together so as to permit the interaction with neighboring particles, the tension was relieved by the migration of particles along the lines of tension, the different parts of atoms migrating in opposite directions. The amount of electricity that passed, then, was clearly related to the chemical affinities of the substances in solution. These experiments led directly to Faraday's two laws of electrochemistry which state:

- The amount of a substance deposited on each electrode of an electrolytic cell is directly proportional to the amount of electricity passing through the cell.
- The quantities of different elements deposited by a given amount of electricity are in the ratio of their chemical equivalent weights.

William Sturgeon built an electric motor in 1832 and invented the commutator, a ring of metal-bristled brushes which allow the spinning armature to maintain contact with the electric current and changed the alternating current to a pulsating direct current. He also improved the voltaic battery and worked on the theory of thermoelectricity.

Hippolyte Pixii, a French instrument maker, constructed the first dynamo in 1832 and later built a direct current dynamo using the commutator. This was the first practical mechanical generator of electric current that used concepts demonstrated by Faraday.

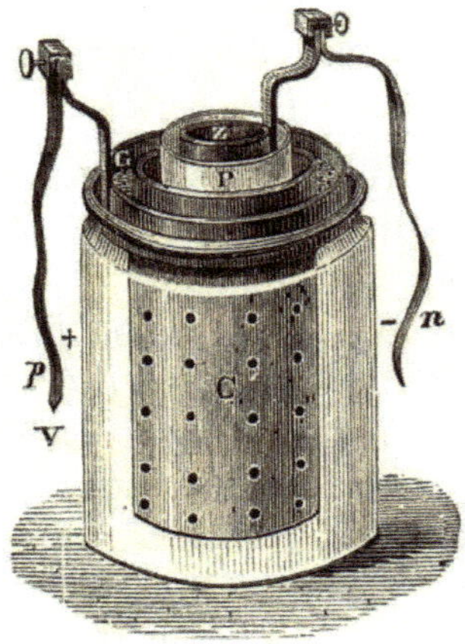

Daniell cell.

John Daniell began experiments in 1835 in an attempt to improve the voltaic battery with its problems of being unsteady and a weak source of electric current. His experiments soon led to remarkable results. In 1836, he invented a primary cell in which hydrogen was eliminated in the generation of the electricity. Daniell had solved the problem of polarization. In his laboratory he had learned to alloy the amalgamated zinc of Sturgeon with mercury. His version was the first of the two-fluid class battery and the first battery that produced a constant reliable source of electric current over a long period of time.

William Grove produced the first fuel cell in 1839. He based his experiment on the fact that sending an electric current through water splits the water into its component parts of hydrogen and oxygen. So, Grove tried reversing the reaction—combining hydrogen and oxygen to produce electricity and water. Eventually the term *fuel cell* was coined in 1889 by Ludwig Mond and Charles Langer, who attempted to build the first practical device using air and industrial coal gas. He also introduced a powerful battery at the annual meeting of the British Association for the Advancement of Science in 1839. Grove's first cell consisted of zinc in diluted sulfuric acid and platinum in concentrated nitric acid, separated by a porous pot. The cell was able to generate about 12 amperes of current at about 1.8 volts. This cell had nearly double the voltage of the first Daniell cell. Grove's nitric acid cell was the favorite battery of the early American telegraph (1840–1860), because it offered strong current output.

As telegraph traffic increased, it was found that the Grove cell discharged poisonous nitrogen dioxide gas. As telegraphs became more complex, the need for a constant voltage became critical and the Grove device was limited (as the cell discharged, nitric acid was depleted and voltage was reduced). By the time of the American Civil War, Grove's battery had been replaced by the Daniell battery. In 1841 Robert Bunsen replaced the expensive platinum electrode used in Grove's battery with a carbon electrode. This led to large scale use of the "Bunsen battery" in the production of arc-lighting and in electroplating.

Wilhelm Weber developed, in 1846, the electrodynamometer, in which a current causes a coil suspended within another coil to turn when a current is passed through both.

In 1852, Weber defined the absolute unit of electrical resistance (which was named the ohm after Georg Ohm). Weber's name is now used as a unit name to describe magnetic flux, the weber.

German physicist Johann Hittorf concluded that *ion movement* caused electric current. In 1853 Hittorf noticed that some ions traveled more rapidly than others. This observation led to the concept of transport number, the rate at which particular ions carried the electric current. Hittorf measured the changes in the concentration of electrolyzed solutions, computed from these the transport numbers (relative carrying capacities) of many ions, and, in 1869, published his findings governing the migration of ions.

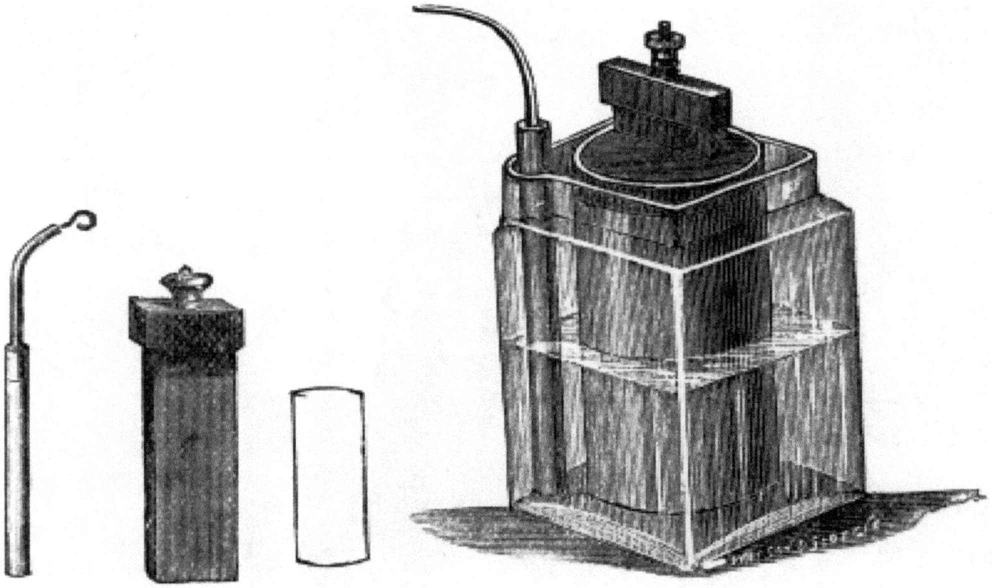

Leclanché cell.

In 1866, Georges Leclanché patented a new battery system, which was immediately successful. Leclanché's original cell was assembled in a porous pot. The positive electrode (the cathode) consisted of crushed manganese dioxide with a little carbon mixed in. The negative pole (anode) was a zinc rod. The cathode was packed into the pot, and a carbon rod was inserted to act as a current collector. The anode and the pot were then immersed in an ammonium chloride solution. The liquid acted as the electrolyte, readily seeping through the porous pot and making contact with the cathode material. Leclanché's "wet" cell became the forerunner to the world's first widely used battery, the zinc-carbon cell.

Late 19th Century Advances and the Advent of Electrochemical Societies

In 1869 Zénobe Gramme devised his first clean direct current dynamo. His generator featured a ring armature wound with many individual coils of wire.

Svante August Arrhenius published his thesis in 1884, *Recherches sur la conductibilité galvanique des électrolytes* (Investigations on the galvanic conductivity of electrolytes). From the results of his experiments, the author concluded that electrolytes, when dissolved in water, become to varying degrees split or dissociated into positive and negative ions. The degree to which this dissociation occurred depended above all on the nature of the substance and its concentration in the solution, being more developed the greater the dilution. The ions were supposed to be the carriers of not only the electric current, as in electrolysis, but also of the chemical activity. The relation between the actual number of ions and their number at great dilution (when all the molecules were dissociated) gave a quantity of special interest ("activity constant").

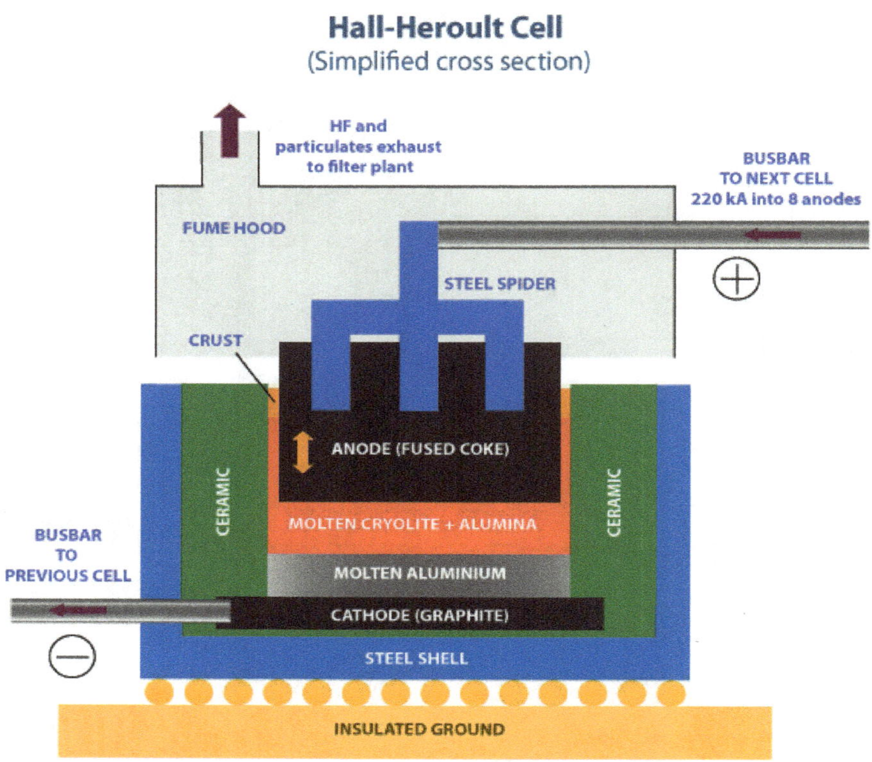

A Hall-Héroult industrial cell.

The race for the commercially viable production of aluminum was won in 1886 by Paul Héroult and Charles M. Hall. The problem many researchers had with extracting aluminum was that electrolysis of an aluminum salt dissolved in water yields aluminum hydroxide. Both Hall and Héroult avoided this problem by dissolving aluminum oxide in a new solvent— fused cryolite (Na_3AlF_6).

Wilhelm Ostwald, 1909 Nobel Laureate, started his experimental work in 1875, with an investigation on the law of mass action of water in relation to the problems of chemical affinity, with special emphasis on electrochemistry and chemical dynamics. In 1894 he gave the first modern definition of a catalyst and turned his attention to catalytic reactions. Ostwald is especially known for his contributions to the field of electrochemistry, including important studies of the electrical conductivity and electrolytic dissociation of organic acids.

Hermann Nernst developed the theory of the electromotive force of the voltaic cell in 1888. He developed methods for measuring dielectric constants and was the first to show that solvents of high dielectric constants promote the ionization of substances. Nernst's early studies in electrochemistry were inspired by Arrhenius' dissociation theory which first recognized the importance of ions in solution. In 1889, Nernst elucidated the theory of galvanic cells by assuming an "electrolytic pressure of dissolution," which forces ions from electrodes into solution and which was opposed to the osmotic pressure of the dissolved ions. He applied the principles of thermodynamics to the chemical reactions proceeding in a battery. In that same year he showed how the characteristics of the current produced could be used to calculate the free energy change in the chemical reaction producing the current. He constructed an equation, known as Nernst Equation, which describes the relation of a battery cell's voltage to its properties.

In 1898 Fritz Haber published his textbook, *Electrochemistry: Grundriss der technischen Elektrochemie auf theoretischer Grundlage* (The Theoretical Basis of Technical Electrochemistry), which was based on the lectures he gave at Karlsruhe. In the preface to his book he expressed his intention to relate chemical research to industrial processes and in the same year he reported the results of his work on electrolytic oxidation and reduction, in which he showed that definite reduction products can result if the voltage at the cathode is kept constant. In 1898 he explained the reduction of nitrobenzene in stages at the cathode and this became the model for other similar reduction processes.

In 1909, Robert Andrews Millikan began a series of experiments to determine the electric charge carried by a single electron. He began by measuring the course of charged water droplets in an electrical field. The results suggested that the charge on the droplets is a multiple of the elementary electric charge, but the experiment was not accurate enough to be convincing. He obtained more precise results in 1910 with his famous oil-drop experiment in which he replaced water (which tended to evaporate too quickly) with oil.

Jaroslav Heyrovský, a Nobel laureate, eliminated the tedious weighing required by previous analytical techniques, which used the differential precipitation of mercury by measuring drop-time. In the previous method, a voltage was applied to a dropping mercury electrode and a reference electrode was immersed in a test solution. After 50 drops of mercury were collected, they were dried and weighed. The applied voltage was varied and the experiment repeated. Measured weight was plotted versus applied voltage to obtain the curve. In 1921, Heyrovský had the idea of measuring the current flowing through the cell instead of just studying drop-time.

On February 10, 1922, the "polarograph" was born as Heyrovský recorded the current-voltage curve for a solution of 1 mol/L NaOH. Heyrovský correctly interpreted the current increase between −1.9 and −2.0 V as being due to the deposit of Na^+ ions, forming an amalgam. Shortly thereafter, with his Japanese colleague Masuzo Shikata, he constructed the first instrument for the automatic recording of polarographic curves, which became world famous later as the polarograph.

Heyrovský's Polarograph.

In 1923, Johannes Nicolaus Brønsted and Thomas Martin Lowry published essentially the same theory about how acids and bases behave using electrochemical basis.

The International Society of Electrochemistry (ISE) was founded in 1949, and some years later the first sophisticated electrophoretic apparatus was developed in 1937 by Arne Tiselius, who was awarded the 1948 Nobel prize for his work in protein electrophoresis. He developed the "moving boundary," which later would become known as *zone electrophoresis*, and used it to separate serum proteins in solution. Electrophoresis became widely developed in the 1940s and 1950s when the technique was applied to molecules ranging from the largest proteins to amino acids and even inorganic ions.

During the 1960s and 1970s quantum electrochemistry was developed by Revaz Dogonadze and his pupils.

PRINCIPLES

Oxidation and Reduction

The term "redox" stands for reduction-oxidation. It refers to electrochemical processes involving electron transfer to or from a molecule or ion changing its oxidation state. This reaction can occur through the application of an external voltage or through the release of chemical energy. Oxidation and reduction describe the change of oxidation state that takes place in the atoms, ions or molecules involved in an electrochemical reaction. Formally, oxidation state is the hypothetical charge that an atom would have if all bonds to atoms of different elements were 100% ionic. An atom or ion that gives up an electron to another atom or ion has its oxidation state increase, and the recipient of the negatively charged electron has its oxidation state decrease.

For example, when atomic sodium reacts with atomic chlorine, sodium donates one electron and attains an oxidation state of +1. Chlorine accepts the electron and its oxidation state is reduced to −1. The sign of the oxidation state (positive/negative) actually corresponds to the value of each ion's electronic charge. The attraction of the differently charged sodium and chlorine ions is the reason they then form an ionic bond.

The loss of electrons from an atom or molecule is called oxidation, and the gain of electrons is reduction. This can be easily remembered through the use of mnemonic devices. Two of the most popular are *"OIL RIG"* (Oxidation Is Loss, Reduction Is Gain) and *"LEO"* the lion says *"GER"* (Lose Electrons: Oxidation, Gain Electrons: Reduction). Oxidation and reduction always occur in a paired fashion such that one species is oxidized when another is reduced. For cases where electrons are shared (covalent bonds) between atoms with large differences in electronegativity, the electron is assigned to the atom with the largest electronegativity in determining the oxidation state.

The atom or molecule which loses electrons is known as the *reducing agent*, or *reductant*, and the substance which accepts the electrons is called the *oxidizing agent*, or *oxidant*. Thus, the oxidizing agent is always being reduced in a reaction; the reducing agent is always being oxidized. Oxygen is a common oxidizing agent, but not the only one. Despite the name, an oxidation reaction does not necessarily need to involve oxygen. In fact, a fire can be fed by an oxidant other than oxygen; fluorine fires are often unquenchable, as fluorine is an even stronger oxidant (it has a higher electronegativity and thus accepts electrons even better) than oxygen.

For reactions involving oxygen, the gain of oxygen implies the oxidation of the atom or molecule to which the oxygen is added (and the oxygen is reduced). In organic compounds, such as butane or ethanol, the loss of hydrogen implies oxidation of the molecule from which it is lost (and the hydrogen is reduced). This follows because the hydrogen donates its electron in covalent bonds with non-metals but it takes the electron along when it is lost. Conversely, loss of oxygen or gain of hydrogen implies reduction.

Balancing Redox Reactions

Electrochemical reactions in water are better understood by balancing redox reactions using the ion-electron method where H^+, OH^- ion, H_2O and electrons (to compensate the oxidation changes) are added to cell's half-reactions for oxidation and reduction.

Acidic Medium

In acid medium H^+ ions and water are added to half-reactions to balance the overall reaction. For example, when manganese reacts with sodium bismuthate.

Unbalanced reaction: $Mn^{2+}_{(aq)} + NaBiO_{3(s)} \rightarrow Bi^{3+}_{(aq)} + MnO_{4\ (aq)}^{-}$

Oxidation: $4\ H_2O_{(l)} + Mn^{2+}_{(aq)} \rightarrow MnO_{4\ (aq)}^{-} + 8\ H^+_{(aq)} + 5\ e^-$

Reduction: $2\ e^- + 6\ H^+_{(aq)} + BiO_3^-{}_{(s)} \rightarrow Bi^{3+}_{(aq)} + 3\ H_2O_{(l)}$

Finally, the reaction is balanced by multiplying the number of electrons from the reduction half reaction to oxidation half reaction and vice versa and adding both half reactions, thus solving the equation.

$8\ H_2O_{(l)} + 2\ Mn^{2+}_{(aq)} \rightarrow 2\ MnO_4^-{}_{(aq)} + 16\ H^+_{(aq)} + 10\ e^-$

$10\ e^- + 30\ H^+_{(aq)} + 5\ BiO_3^-{}_{(s)} \rightarrow 5\ Bi^{3+}_{(aq)} + 15\ H_2O_{(l)}$

Reaction balanced:

$14\ H^+_{(aq)} + 2\ Mn^{2+}_{(aq)} + 5\ NaBiO_{3(s)} \rightarrow 7\ H_2O_{(l)} + 2\ MnO_4^-{}_{(aq)} + 5\ Bi^{3+}_{(aq)} + 5\ Na^+_{(aq)}$

Basic Medium

In basic medium OH^- ions and water are added to half reactions to balance the overall reaction. For example, on reaction between potassium permanganate and sodium sulfite.

Unbalanced reaction: $KMnO_4 + Na_2SO_3 + H_2O \rightarrow MnO_2 + Na_2SO_4 + KOH$

Reduction: $3\ e^- + 2\ H_2O + MnO_4^- \rightarrow MnO_2 + 4\ OH^-$

Oxidation: $2\ OH^- + SO_3^{2-} \rightarrow SO_4^{2-} + H_2O + 2\ e^-$

The same procedure as followed on acid medium by multiplying electrons to opposite half reactions solve the equation thus balancing the overall reaction.

$6\ e^- + 4\ H_2O + 2\ MnO_4^- \rightarrow 2\ MnO_2 + 8\ OH^-$

$6\ OH^- + 3\ SO_3^{2-} \rightarrow 3\ SO_4^{2-} + 3\ H_2O + 6e^-$

Equation balanced:

$2\ KMnO_4 + 3\ Na_2SO_3 + H_2O \rightarrow 2\ MnO_2 + 3\ Na_2SO_4 + 2\ KOH$

Neutral Medium

The same procedure as used on acid medium is applied, for example on balancing using electron ion method to complete combustion of propane.

Unbalanced reaction: $C_3H_8 + O_2 \rightarrow CO_2 + H_2O$

Reduction: $4\ H^+ + O_2 + 4\ e^- \rightarrow 2\ H_2O$

Oxidation: $6\ H_2O + C_3H_8 \rightarrow 3\ CO_2 + 20\ e^- + 20\ H^+$

As in acid and basic medium, electrons which were used to compensate oxidation changes are multiplied to opposite half reactions, thus solving the equation.

$20\ H^+ + 5\ O_2 + 20\ e^- \rightarrow 10\ H_2O$

$6\ H_2O + C_3H_8 \rightarrow 3\ CO_2 + 20\ e^- + 20\ H^+$

Equation balanced:

$C_3H_8 + 5\ O_2 \rightarrow 3\ CO_2 + 4\ H_2O$

Electrochemical Cells

An electrochemical cell is a device that produces an electric current from energy released by a spontaneous redox reaction. This kind of cell includes the Galvanic cell or Voltaic cell, named after Luigi Galvani and Alessandro Volta, both scientists who conducted several experiments on chemical reactions and electric current during the late 18th century.

Electrochemical cells have two conductive electrodes (the anode and the cathode). The anode is defined as the electrode where oxidation occurs and the cathode is the electrode where the reduction takes place. Electrodes can be made from any sufficiently conductive materials, such as metals, semiconductors, graphite, and even conductive polymers. In between these electrodes is the electrolyte, which contains ions that can freely move.

The galvanic cell uses two different metal electrodes, each in an electrolyte where the positively charged ions are the oxidized form of the electrode metal. One electrode will undergo oxidation (the anode) and the other will undergo reduction (the cathode). The metal of the anode will oxidize, going from an oxidation state of 0 (in the solid form) to a positive oxidation state and become an ion. At the cathode, the metal ion in solution will accept one or more electrons from the cathode and the ion's oxidation state is reduced to 0. This forms a solid metal that electrodeposits on the cathode. The two electrodes must be electrically connected to each other, allowing for a flow of electrons that leave the metal of the anode and flow through this connection to the ions at the surface of the cathode. This flow of electrons is an electric current that can be used to do work, such as turn a motor or power a light.

A galvanic cell whose electrodes are zinc and copper submerged in zinc sulfate and copper sulfate, respectively, is known as a Daniell cell.

Half reactions for a Daniell cell are these:

Zinc electrode (anode): $Zn_{(s)} \rightarrow Zn^{2+}_{(aq)} + 2\ e^-$

Copper electrode (cathode): $Cu^{2+}_{(aq)} + 2\ e^- \rightarrow Cu_{(s)}$

A modern cell stand for electrochemical research. The electrodes attach to high-quality metallic wires, and the stand is attached to a potentiostat/galvanostat (not pictured). A shot glass-shaped container is aerated with a noble gas and sealed with the Teflon block.

In this example, the anode is the zinc metal which is oxidized (loses electrons) to form zinc ions in solution, and copper ions accept electrons from the copper metal electrode and the ions deposit at the copper cathode as an electrodeposit. This cell forms a simple battery as it will spontaneously generate a flow of electric current from the anode to the cathode through the external connection. This reaction can be driven in reverse by applying a voltage, resulting in the deposition of zinc metal at the anode and formation of copper ions at the cathode.

To provide a complete electric circuit, there must also be an ionic conduction path between the anode and cathode electrolytes in addition to the electron conduction path. The simplest ionic conduction path is to provide a liquid junction. To avoid mixing between the two electrolytes, the liquid junction can be provided through a porous plug that allows ion flow while reducing electrolyte mixing. To further minimize mixing of the electrolytes, a salt bridge can be used which consists of an electrolyte saturated gel in an inverted U-tube. As the negatively charged electrons flow in one direction around this circuit, the positively charged metal ions flow in the opposite direction in the electrolyte.

A voltmeter is capable of measuring the change of electrical potential between the anode and the cathode.

Electrochemical cell voltage is also referred to as electromotive force or emf.

A cell diagram can be used to trace the path of the electrons in the electrochemical cell. For example, here is a cell diagram of a Daniell cell:

$$Zn_{(s)} \mid Zn^{2+}\ (1M) \parallel Cu^{2+}\ (1M) \mid Cu_{(s)}$$

First, the reduced form of the metal to be oxidized at the anode (Zn) is written. This is separated from its oxidized form by a vertical line, which represents the limit between the phases (oxidation changes). The double vertical lines represent the saline bridge on the cell. Finally, the oxidized form of the metal to be reduced at the cathode, is written, separated from its reduced form by the vertical line. The electrolyte concentration is given as it is an important variable in determining the cell potential.

Standard Electrode Potential

To allow prediction of the cell potential, tabulations of standard electrode potential are available. Such tabulations are referenced to the standard hydrogen electrode (SHE). The standard hydrogen electrode undergoes the reaction

$$2\ H^{+}_{(aq)} + 2\ e^{-} \rightarrow H_2$$

which is shown as reduction but, in fact, the SHE can act as either the anode or the cathode, depending on the relative oxidation/reduction potential of the other electrode/ electrolyte combination. The term standard in SHE requires a supply of hydrogen gas bubbled through the electrolyte at a pressure of 1 atm and an acidic electrolyte with H^+ activity equal to 1 (usually assumed to be $[H^+] = 1$ mol/liter).

The SHE electrode can be connected to any other electrode by a salt bridge to form a cell. If the second electrode is also at standard conditions, then the measured cell potential is called the standard electrode potential for the electrode. The standard electrode potential for the SHE is zero, by definition. The polarity of the standard electrode potential provides information about the relative reduction potential of the electrode compared to the SHE. If the electrode has a positive potential with respect to the SHE, then that means it is a strongly reducing electrode which forces the SHE to be the anode (an example is Cu in aqueous $CuSO_4$ with a standard electrode potential of 0.337 V). Conversely, if the measured potential is negative, the electrode is more oxidizing than the SHE (such as Zn in $ZnSO_4$ where the standard electrode potential is −0.76 V).

Standard electrode potentials are usually tabulated as reduction potentials. However, the reactions are reversible and the role of a particular electrode in a cell depends on the relative oxidation/reduction potential of both electrodes. The oxidation potential for a particular electrode is just the negative of the reduction potential. A standard cell potential can be determined by looking up the standard electrode potentials for both electrodes (sometimes called half cell potentials). The one that is smaller will be the anode and will undergo oxidation. The cell potential is then calculated as the sum of the reduction potential for the cathode and the oxidation potential for the anode.

$$E^\circ_{cell} = E^\circ_{red}\text{ (cathode)} - E^\circ_{red}\text{ (anode)} = E^\circ_{red}\text{ (cathode)} + E^\circ_{oxi}\text{ (anode)}$$

For example, the standard electrode potential for a copper electrode is:

Cell diagram

$Pt_{(s)} \mid H_2\ (1\ atm) \mid H^+\ (1\ M) \parallel Cu^{2+}\ (1\ M) \mid Cu_{(s)}$

$$E^\circ_{cell} = E^\circ_{red}\text{ (cathode)} - E^\circ_{red}\text{ (anode)}$$

At standard temperature, pressure and concentration conditions, the cell's emf (measured by a multimeter) is 0.34 V. By definition, the electrode potential for the SHE is zero. Thus, the Cu is the cathode and the SHE is the anode giving

$$E_{cell} = E^\circ(Cu^{2+}/Cu) - E^\circ(H^+/H_2)$$

Or,

$$E^\circ(Cu^{2+}/Cu) = 0.34\ V$$

Changes in the stoichiometric coefficients of a balanced cell equation will not change E°_{red} value because the standard electrode potential is an intensive property.

Spontaneity of Redox Reaction

During operation of electrochemical cells, chemical energy is transformed into electrical energy and is expressed mathematically as the product of the cell's emf and the electric charge transferred through the external circuit.

$$\text{Electrical energy} = E_{cell}C_{trans}$$

where E_{cell} is the cell potential measured in volts (V) and C_{trans} is the cell current integrated over time and measured in coulombs (C); C_{trans} can also be determined by multiplying the total number of electrons transferred (measured in moles) times Faraday's constant (F).

The emf of the cell at zero current is the maximum possible emf. It is used to calculate the maximum possible electrical energy that could be obtained from a chemical reaction. This energy is referred to as electrical work and is expressed by the following equation:

$$W_{max} = W_{electrical} = -nFE_{cell},$$

where work is defined as positive into the system.

Since the free energy is the maximum amount of work that can be extracted from a system, one can write:

$$\Delta G = -nFE_{cell}$$

A positive cell potential gives a negative change in Gibbs free energy. This is consistent with the cell production of an electric current from the cathode to the anode through the external circuit. If the current is driven in the opposite direction by imposing an external potential, then work is done on the cell to drive electrolysis.

A spontaneous electrochemical reaction (change in Gibbs free energy less than zero) can be used to generate an electric current in electrochemical cells. This is the basis of all batteries and fuel cells. For example, gaseous oxygen (O_2) and hydrogen (H_2) can be combined in a fuel cell to form water and energy, typically a combination of heat and electrical energy.

Conversely, non-spontaneous electrochemical reactions can be driven forward by the application of a current at sufficient voltage. The electrolysis of water into gaseous oxygen and hydrogen is a typical example.

The relation between the equilibrium constant, *K*, and the Gibbs free energy for an electrochemical cell is expressed as follows:

$$\Delta G^\circ = -RT \ln K = -nFE^\circ_{cell}.$$

Rearranging to express the relation between standard potential and equilibrium constant yields

$$E^o_{cell} = \frac{RT}{nF} \ln K.$$

The previous equation can use Briggsian logarithm as shown below:

$$E^o_{cell} = \frac{0.0591\,\text{V}}{n} \log K$$

Cell emf dependency on changes in concentration

Nernst Equation

The standard potential of an electrochemical cell requires standard conditions (ΔG°) for all of the reactants. When reactant concentrations differ from standard conditions, the cell potential will deviate from the standard potential. In the 20th century German chemist Walther Nernst proposed a mathematical model to determine the effect of reactant concentration on electrochemical cell potential.

In the late 19th century, Josiah Willard Gibbs had formulated a theory to predict whether a chemical reaction is spontaneous based on the free energy

$$\Delta G = \Delta G^\circ + RT \ln Q$$

Here ΔG is change in Gibbs free energy, $\Delta G°$ is the cell potential when Q is equal to 1, T is absolute temperature(Kelvin), R is the gas constant and Q is reaction quotient which can be found by dividing products by reactants using only those products and reactants that are aqueous or gaseous.

Gibbs' key contribution was to formalize the understanding of the effect of reactant concentration on spontaneity.

Based on Gibbs' work, Nernst extended the theory to include the contribution from electric potential on charged species. As shown in the previous section, the change in Gibbs free energy for an electrochemical cell can be related to the cell potential. Thus, Gibbs' theory becomes

$$nF\Delta E = nF\Delta E^{\circ} - RT\ln Q$$

Here n is the number of electrons/mole product, F is the Faraday constant (coulombs/mole), and ΔE is cell potential.

Finally, Nernst divided through by the amount of charge transferred to arrive at a new equation which now bears his name:

$$\Delta E = \Delta E^{\circ} - \frac{RT}{nF}\ln Q$$

Assuming standard conditions (T = 25 °C) and R = 8.3145 J/(K·mol), the equation above can be expressed on base—10 logarithm as shown below:

$$\Delta E = \Delta E^{\circ} - \frac{0.05916\,\text{V}}{n}\log Q$$

Concentration Cells

A concentration cell is an electrochemical cell where the two electrodes are the same material, the electrolytes on the two half-cells involve the same ions, but the electrolyte concentration differs between the two half-cells.

An example is an electrochemical cell, where two copper electrodes are submerged in two copper(II) sulfate solutions, whose concentrations are 0.05 M and 2.0 M, connected through a salt bridge. This type of cell will generate a potential that can be predicted by the Nernst equation. Both can undergo the same chemistry (although the reaction proceeds in reverse at the anode)

$Cu^{2+}_{(aq)} + 2\ e^{-} \rightarrow Cu_{(s)}$

Le Chatelier's principle indicates that the reaction is more favorable to reduction as the concentration of Cu^{2+} ions increases. Reduction will take place in the cell's compartment where concentration is higher and oxidation will occur on the more dilute side.

The following cell diagram describes the cell mentioned above:

$Cu_{(s)}\ |\ Cu^{2+}\ (0.05\ M)\ ||\ Cu^{2+}\ (2.0\ M)\ |\ Cu_{(s)}$

Where the half cell reactions for oxidation and reduction are:

Oxidation: $Cu_{(s)} \rightarrow Cu^{2+}\ (0.05\ M) + 2\ e^{-}$

Reduction: Cu^{2+} (2.0 M) + 2 e^- → $Cu_{(s)}$

Overall reaction: Cu^{2+} (2.0 M) → Cu^{2+} (0.05 M)

The cell's emf is calculated through Nernst equation as follows:

$$E = E^{\circ} - \frac{0.05916\,\text{V}}{2} \log \frac{[\text{Cu}^{2+}]_{diluted}}{[\text{Cu}^{2+}]_{concentrated}}$$

The value of E° in this kind of cell is zero, as electrodes and ions are the same in both half-cells.

After replacing values from the case mentioned, it is possible to calculate cell›s potential:

$$E = 0 - \frac{0.05916\,\text{V}}{2} \log \frac{0.05}{2.0} = 0.0474\,\text{V}$$

or by:

$$E = 0 - \frac{0.0257\,\text{V}}{2} \ln \frac{0.05}{2.0} = 0.0474\,\text{V}$$

However, this value is only approximate, as reaction quotient is defined in terms of ion activities which can be approximated with the concentrations as calculated here.

The Nernst equation plays an important role in understanding electrical effects in cells and organelles. Such effects include nerve synapses and cardiac beat as well as the resting potential of a somatic cell.

BATTERY

Many types of battery have been commercialized and represent an important practical application of electrochemistry. Early wet cells powered the first telegraph and telephone systems, and were the source of current for electroplating. The zinc-manganese dioxide dry cell was the first portable, non-spillable battery type that made flashlights and other portable devices practical. The mercury battery using zinc and mercuric oxide provided higher levels of power and capacity than the original dry cell for early electronic devices, but has been phased out of common use due to the danger of mercury pollution from discarded cells.

The lead–acid battery was the first practical secondary (rechargeable) battery that could have its capacity replenished from an external source. The electrochemical reaction that produced current was (to a useful degree) reversible, allowing electrical energy and chemical energy to be interchanged as needed. Common lead acid batteries contain a mixture of acid and water, as well as lead plates. The most common mixture used today is 30% acid. One problem however is if left uncharged acid will crystallize within the lead plates of the battery rendering it useless. These batteries last an average of 3 years with daily use however it is not unheard of for a lead acid battery to still be functional after 7–10 years. Lead-acid cells continue to be widely used in automobiles.

All the preceding types have water-based electrolytes, which limits the maximum voltage per cell. The freezing of water limits low temperature performance. The lithium battery, which does not (and cannot) use water in the electrolyte, provides improved performance over other types; a rechargeable lithium ion battery is an essential part of many mobile devices.

The flow battery, an experimental type, offers the option of vastly larger energy capacity because its reactants can be replenished from external reservoirs. The fuel cell can turn the chemical energy bound in hydrocarbon gases or hydrogen directly into electrical energy with much higher efficiency than any combustion process; such devices have powered many spacecraft and are being applied to grid energy storage for the public power system.

CORROSION

Corrosion is an electrochemical process, which reveals itself in rust or tarnish on metals like iron or copper and their respective alloys, steel and brass.

Iron Corrosion

For iron rust to occur the metal has to be in contact with oxygen and water, although chemical reactions for this process are relatively complex and not all of them are completely understood. It is believed the causes are the following: Electron transfer (reduction-oxidation)

One area on the surface of the metal acts as the anode, which is where the oxidation (corrosion) occurs. At the anode, the metal gives up electrons.

$$Fe_{(s)} \rightarrow Fe^{2+}_{(aq)} + 2\ e^-$$

Electrons are transferred from iron, reducing oxygen in the atmosphere into water on the cathode, which is placed in another region of the metal.

$$O_{2(g)} + 4\ H^+_{(aq)} + 4\ e^- \rightarrow 2\ H_2O_{(l)}$$

Global reaction for the process:

$$2\ Fe_{(s)} + O_{2(g)} + 4\ H^+_{(aq)} \rightarrow 2\ Fe^{2+}_{(aq)} + 2\ H_2O_{(l)}$$

Standard emf for iron rusting:

$$E° = E°\ (cathode) - E°\ (anode)$$

$$E° = 1.23V - (-0.44\ V) = 1.67\ V$$

Iron corrosion takes place in an acid medium; H^+ ions come from reaction between carbon dioxide in the atmosphere and water, forming carbonic acid. Fe^{2+} ions oxidizes, following this equation:

$$4\,Fe^{2+}_{(aq)} + O_{2(g)} + (4 + 2\,x)H_2O_{(l)} \longrightarrow 2\,Fe_2O_3 \cdot xH_2O + 8\,H^+_{(aq)}$$

Iron(III) oxide hydrate is known as rust. The concentration of water associated with iron oxide varies, thus the chemical formula is represented by $2\,Fe_2O_3 \cdot xH_2O$.

An electric circuit is formed as passage of electrons and ions occurs, thus if an electrolyte is present it will facilitate oxidation, explaining why rusting is quicker in salt water.

Corrosion of Common Metals

Coinage metals, such as copper and silver, slowly corrode through use. A patina of green-blue copper carbonate forms on the surface of copper with exposure to the water and carbon dioxide in the air. Silver coins or cutlery that are exposed to high sulfur foods such as eggs or the low levels of sulfur species in the air develop a layer of black Silver sulfide.

Gold and platinum are extremely difficult to oxidize under normal circumstances, and require exposure to a powerful chemical oxidizing agent such as aqua regia.

Some common metals oxidize extremely rapidly in air. Titanium and aluminum oxidize instantaneously in contact with the oxygen in the air. These metals form an extremely thin layer of oxidized metal on the surface. This thin layer of oxide protects the underlying layers of the metal from the air preventing the entire metal from oxidizing. These metals are used in applications where corrosion resistance is important. Iron, in contrast, has an oxide that forms in air and water, called rust, that does not stop the further oxidation of the iron. Thus iron left exposed to air and water will continue to rust until all of the iron is oxided.

Prevention of Corrosion

Attempts to save a metal from becoming anodic are of two general types. Anodic regions dissolve and destroy the structural integrity of the metal.

While it is almost impossible to prevent anode/cathode formation, if a non-conducting material covers the metal, contact with the electrolyte is not possible and corrosion will not occur.

Coating

Metals can be coated with paint or other less conductive metals (*passivation*). This prevents the metal surface from being exposed to electrolytes. Scratches exposing the metal substrate will result in corrosion. The region under the coating adjacent to the scratch acts as the anode of the reaction.

Sacrificial Anodes

A method commonly used to protect a structural metal is to attach a metal which is more anodic than the metal to be protected. This forces the structural metal to be cathodic, thus spared corrosion. It is called *"sacrificial"* because the anode dissolves and has to be replaced periodically.

Zinc bars are attached to various locations on steel ship hulls to render the ship hull cathodic. The zinc bars are replaced periodically. Other metals, such as magnesium, would work very well but zinc is the least expensive useful metal.

To protect pipelines, an ingot of buried or exposed magnesium (or zinc) is buried beside the pipeline and is connected electrically to the pipe above ground. The pipeline is forced to be a cathode and is protected from being oxidized and rusting. The magnesium anode is sacrificed. At intervals new ingots are buried to replace those lost.

Electrolysis

The spontaneous redox reactions of a conventional battery produce electricity through the different chemical potentials of the cathode and anode in the electrolyte. However, electrolysis requires an external source of electrical energy to induce a chemical reaction, and this process takes place in a compartment called an electrolytic cell.

Electrolysis of Molten Sodium Chloride

When molten, the salt sodium chloride can be electrolyzed to yield metallic sodium and gaseous chlorine. Industrially this process takes place in a special cell named Down's cell. The cell is connected to an electrical power supply, allowing electrons to migrate from the power supply to the electrolytic cell.

Reactions that take place at Down's cell are the following:

Anode (oxidation): $2\ Cl^- \rightarrow Cl_{2(g)} + 2\ e^-$

Cathode (reduction): $2\ Na^+_{(l)} + 2\ e^- \rightarrow 2\ Na_{(l)}$

Overall reaction: $2\ Na^+ + 2\ Cl^-_{(l)} \rightarrow 2\ Na_{(l)} + Cl_{2(g)}$

This process can yield large amounts of metallic sodium and gaseous chlorine, and is widely used on mineral dressing and metallurgy industries.

The emf for this process is approximately −4 V indicating a (very) non-spontaneous process. In order for this reaction to occur the power supply should provide at least a potential of 4 V. However, larger voltages must be used for this reaction to occur at a high rate.

Electrolysis of Water

Water can be converted to its component elemental gasses, H_2 and O_2 through the application of an external voltage. Water doesn't decompose into hydrogen and oxygen spontaneously as the Gibbs free energy for the process at standard conditions is about 474.4 kJ. The decomposition of water into hydrogen and oxygen can be performed in an electrolytic cell. In it, a pair of inert electrodes usually made of platinum immersed in water act as anode and cathode in the electrolytic process. The electrolysis starts with the application of an external voltage between the electrodes. This process will not occur except at extremely high voltages without an electrolyte such as sodium chloride or sulfuric acid (most used 0.1 M).

Bubbles from the gases will be seen near both electrodes. The following half reactions describe the process mentioned above:

Anode (oxidation): $2\ H_2O_{(l)} \rightarrow O_{2(g)} + 4\ H^+_{(aq)} + 4\ e^-$

Cathode (reduction): $2\ H_2O_{(g)} + 2\ e^- \rightarrow H_{2(g)} + 2\ OH^-_{(aq)}$

Overall reaction: $2\ H_2O_{(l)} \rightarrow 2\ H_{2(g)} + O_{2(g)}$

Although strong acids may be used in the apparatus, the reaction will not net consume the acid. While this reaction will work at any conductive electrode at a sufficiently large potential, platinum catalyzes both hydrogen and oxygen formation, allowing for relatively mild voltages (~2 V depending on the pH).

Electrolysis of Aqueous Solutions

Electrolysis in an aqueous is a similar process as mentioned in electrolysis of water. However, it is considered to be a complex process because the contents in solution have to be analyzed in half reactions, whether reduced or oxidized.

Electrolysis of a Solution of Sodium Chloride

The presence of water in a solution of sodium chloride must be examined in respect to its reduction and oxidation in both electrodes. Usually, water is electrolyzed as mentioned in electrolysis of water yielding *gaseous oxygen in the anode* and gaseous hydrogen in the cathode. On the other hand, sodium chloride in water dissociates in Na^+ and Cl^- ions, cation, which is the positive ion, will be attracted to the cathode (-), thus reducing the sodium ion. The anion will then be attracted to the anode (+) oxidizing chloride ion.

The following half reactions describes the process mentioned:

1. Cathode: $Na^+_{(aq)} + e^- \rightarrow Na_{(s)}$ $E°_{red} = -2.71$ V
2. Anode: $2\ Cl^-_{(aq)} \rightarrow Cl_{2(g)} + 2\ e^-$ $E°_{red} = +1.36$ V
3. Cathode: $2\ H_2O_{(l)} + 2\ e^- \rightarrow H_{2(g)} + 2\ OH^-_{(aq)}$ $E°_{red} = -0.83$ V
4. Anode: $2\ H_2O_{(l)} \rightarrow O_{2(g)} + 4\ H^+_{(aq)} + 4\ e^-$ $E°_{red} = +1.23$ V

Reaction 1 is discarded as it has the most negative value on standard reduction potential thus making it less thermodynamically favorable in the process.

When comparing the reduction potentials in reactions 2 and 4, the reduction of chloride ion is favored. Thus, if the Cl^- ion is favored for reduction, then the water reaction is favored for oxidation producing gaseous oxygen, however experiments show gaseous chlorine is produced and not oxygen.

Although the initial analysis is correct, there is another effect that can happen, known as the overvoltage effect. Additional voltage is sometimes required, beyond the voltage predicted by the $E°_{cell}$. This may be due to kinetic rather than thermodynamic considerations. In fact, it has been proven that the activation energy for the chloride ion is very low, hence favorable in kinetic terms. In other words, although the voltage applied is thermodynamically sufficient to drive electrolysis, the rate is so slow that to make the process proceed in a reasonable time frame, the voltage of the external source has to be increased (hence, overvoltage).

Finally, reaction 3 is favorable because it describes the proliferation of OH^- ions thus letting a probable reduction of H^+ ions less favorable an option.

The overall reaction for the process according to the analysis would be the following:

Anode (oxidation): $2\ Cl^-_{(aq)} \rightarrow Cl_{2(g)} + 2\ e^-$

Cathode (reduction): $2\ H_2O_{(l)} + 2\ e^- \rightarrow H_{2(g)} + 2\ OH^-_{(aq)}$

Overall reaction: $2\ H_2O + 2\ Cl^-_{(aq)} \rightarrow H_{2(g)} + Cl_{2(g)} + 2\ OH^-_{(aq)}$

As the overall reaction indicates, the concentration of chloride ions is reduced in comparison to OH^- ions (whose concentration increases). The reaction also shows the production of gaseous hydrogen, chlorine and aqueous sodium hydroxide.

Quantitative Electrolysis and Faraday's Laws

Quantitative aspects of electrolysis were originally developed by Michael Faraday in 1834. Faraday is also credited to have coined the terms *electrolyte*, electrolysis, among many others while he studied quantitative analysis of electrochemical reactions. Also he was an advocate of the law of conservation of energy.

First Law

Faraday concluded after several experiments on electric current in non-spontaneous process, the mass of the products yielded on the electrodes was proportional to the value of current supplied to the cell, the length of time the current existed, and the molar mass of the substance analyzed. In other words, the amount of a substance deposited on each electrode of an electrolytic cell is directly proportional to the quantity of electricity passed through the cell.

Below is a simplified equation of Faraday's first law:

$$m = \frac{1}{96485(\mathrm{C} \cdot \mathrm{mol}^{-1})} \cdot \frac{QM}{n}$$

Where

m is the mass of the substance produced at the electrode (in grams),

Q is the total electric charge that passed through the solution (in coulombs),

n is the valence number of the substance as an ion in solution (electrons per ion),

M is the molar mass of the substance (in grams per mole).

Second Law

Faraday devised the laws of chemical electrodeposition of metals from solutions in 1857. He formulated the second law of electrolysis stating *"the amounts of bodies which are equivalent to each other in their ordinary chemical action have equal quantities of electricity naturally associated with them."* In other words, the quantities of different elements deposited by a given amount of electricity are in the ratio of their chemical equivalent weights.

An important aspect of the second law of electrolysis is electroplating which together with the first law of electrolysis, has a significant number of applications in the industry, as when used to protect metals to avoid corrosion.

APPLICATIONS

There are various extremely important electrochemical processes in both nature and industry, like the coating of objects with metals or metal oxides through electrodeposition and the detection of alcohol in drunken drivers through the redox reaction of ethanol. The generation of chemical energy through photosynthesis is inherently an electrochemical process, as is production of metals like aluminum and titanium from their ores. Certain diabetes blood sugar meters measure the amount of glucose in the blood through its redox potential. As well as the established electrochemical technologies (like deep cycle lead acid batteries) there is also a wide range of new emerging technologies such as fuel cells, large format lithium ion batteries, electrochemical reactors and super-capacitors that are becoming increasingly commercial. Electrochemistry has also important applications in the food industry,like the assessment of food/package interactions, the analysis of milk composition, the characterization and the determination of the freezing end-point of ice-cream mixes, the determination of free acidity in olive oil.

The action potentials that travel down neurons are based on electric current generated by the movement of sodium and potassium ions into and out of cells. Specialized cells in certain animals like the electric eel can generate electric currents powerful enough to disable much larger animals.

Electrochemical Engineering

Electrochemical engineering is the branch of engineering dealing with the technological applications of electrochemical phenomena, such as synthesis of chemicals, electrowinning and refining of metals, batteries and fuel cells, sensors, surface modification by electrodeposition and etching, separations, and corrosion. It is an overlap between electrochemistry and chemical engineering. One of the pioneers of this field of engineering was Charles Frederick Burgess.

More than 6% of the electrical energy is consumed by electrochemical operations in the USA.

History

This branch of engineering emerged gradually from chemical engineering. The works of Wagner (1962) and Levich (1962) influenced the emergence of electrochemical engineering, because their work inspired so many others. Several individuals, including Tobias, Ibl, and Hine, established engineering training centers and, with their colleagues, developed important experimental and theoretical methods of study.

Scope

Electrochemical engineering combines the study of heterogeneous charge transfer at electrode/electrolyte interphases with the development of practical materials and processes. Fundamental considerations include electrode materials and the kinetics of redox species. The development of the technology involves the study of the electrochemical reactors, their potential and current distribution, mass transport conditions, hydrodynamics, geometry and components as well as the quantification of its overall performance in terms of reaction yield, conversion efficiency, and energy efficiency. Industrial developments require further reactor and process design, fabrication methods, testing and product development.

Applications

Electrochemical engineering finds applications in chemical synthesis, ion/organics removal, deposition of fils of metals and semiconductors, sensors and monitoring and energy storage and conversion.

REFERENCES

1. Richard P. Olenick, Tom M. Apostol, David L. Goodstein Beyond the mechanical universe: from electricity to modern physics, Cambridge University Press (1986) ISBN 0-521-30430-X, p. 160
2. R. Hellborg Electrostatic accelerators: fundamentals and applications (2005) ISBN 3540239839 p. 52
3. Steven Weinberg The discovery of subatomic particles Cambridge University Press (2003) ISBN 0-521-82351-X, p. 15
4. J. A. M. Bleeker, Johannes Geiss, M. Huber The century of space science, Volume 1, Springer (2001) ISBN 0-7923-7196-8 p. 227
5. John Robert Norris, Douglas W. Ribbons (1972) Methods in microbiology, Volume 6, Academic Press. ISBN 0-12-521546-0 p. 248
6. Frederick Collier Bakewell Electric science; its history, phenomena, and applications, Ingram, Cooke (1853) pp. 27–31
7. Keith James Laidler The world of physical chemistry, Oxford University Press (1995) ISBN 0-19-855919-4 pp. 219–220
8. The New Encyclopaedia Britannica: Micropædia, Vol. 10 (1991) ISBN 0-85229-529-4, p. 90
9. Charles Knight (ed.) Biography: or, Third division of "The English encyclopedia", Volume 2, Bradbury, Evans & Co. (1867)
10. William Berkson (1974) Fields of force: the development of a world view from Faraday to Einstein, Routledge. ISBN 0-7100-7626-6 pp. 34 ff
11. Brian Scott Baigrie Electricity and magnetism: a historical perspective, Greenwood Publishing Group (2007) ISBN 0-313-33358-0 p. 73
12. Nobel Lectures, p. 59
13. Polmear, I.J. (2006). "Production of Aluminium". Light alloys from traditional alloys to nanocrystals. Oxford: Elsevier/Butterworth-Heinemann. pp. 15–16. ISBN 978-0-7506-6371-7.
14. Nobel Lectures, p. 170
15. Nobel Lectures, p. 363

16. Nobel Lectures, p. 342
17. ECS History Center
18. Millikan, Robert A. (1911). "The Isolation of an Ion, a Precision Measurement of its Charge, and the Correction of Stokes' Law". Physical Review. 32 (2): 349–397. Bibcode:1911PhRvI..32..349M. doi:10.1103/PhysRevSeriesI.32.349.
19. William L. Masterton, Cecile N. Hurley Chemistry: Principles and Reactions, Cengage Learning (2008) ISBN 0-495-12671-3 p. 379
20. The Nobel Prize in Chemistry 1948 Arne Tiselius, nobelprize.org
21. The International Society of Electrochemistry
22. Wiberg, pp. 215–216
23. Swaddle, pp. 308–314
24. Wiberg, pp. 210–212
25. Badwal, Sukhvinder P. S.; Giddey, Sarbjit S.; Munnings, Christopher; Bhatt, Anand I.; Hollenkamp, Anthony F. (24 September 2014). "Emerging electrochemical energy conversion and storage technologies (open access)". Frontiers in Chemistry. 2. Bibcode:2014FrCh....2...79B. doi:10.3389/fchem.2014.00079. PMC 4174133Freely accessible. PMID 25309898.
26. Ebbing, pp. 800–801
27. Wiberg, pp. 235–239
28. Ebbing, pp. 837–839
29. Wiberg, p. 65
30. Faraday, Michael (1791–1867), Wolfram Research
31. McIntyre, J.M.; Pham, H.Q. (1996). "Electrochemical impedance spectroscopy; a tool for organic coatings optimizations". Progress in Organic Coatings. 27 (1-4): 201–207. doi:10.1016/0300-9440(95)00532-3.
32. Amirudin, A.; Thieny, D. (1995). "Application of electrochemical impedance spectroscopy to study the degradation of polymer-coated metals". Progress in Organic Coatings. 26 (1): 1–28. doi:10.1016/0300-9440(95)00581-1.
33. S.P.S BADWAL; S. Giddey; C. Munnings; A. I. Bhatt; A. Hollenkamp (2014). "Emerging electrochemical energy conversion and storage technologies". Frontiers in Chemistry. 2: 79. Bibcode:2014FrCh....2...79B. doi:10.3389/fchem.2014.00079. PMC 4174133Freely accessible. PMID 25309898.
34. Hollaender, J. (2009). "Rapid assessment of food/package interactions by electrochemical impedance spectroscopy (EIS)". Food Additives & Contaminants. 14 (6-7): 617–626. doi:10.1080/02652039709374574.
35. Mabrook, M.F.; Petty, M.C. (2003). "Effect of composition on the electrical conductance of milk". Journal of Food Engineering. 60 (3): 321–325. doi:10.1016/S0260-8774(03)00054-2.

CHAPTER 2

Electrode Potential

Chapter Contents

Electrode potential for any given electrode is the voltage or potential difference of a cell assembled from a standard hydrogen electrode and the given electrode with the potential that is being defined. It is the resultant potential difference between a point on the electrode surface and the point in the bulk of electrolyte, due to transfer of charged particles as well as the adsorption of some polar molecules.

As per convention, the standard hydrogen electrode is always at zero potential.

Electrode potential has uses such as:

- Prediction of corrosion-related chemical or electrochemical reactions
- Used for choosing substances and devices for controlling reactions
- Helps the study of crevice corrosion and pitting, as electrode potential in crevices and pits is studied for controlling reactions

INTRODUCTION

Electrode potential, *E*, in chemistry or electrochemistry, according to a IUPAC definition, is the electromotive force of a cell built of two electrodes:

- on the left-hand side is the standard hydrogen electrode, and
- on the right-hand side is the electrode the potential of which is being defined.

By convention:

$$E_{\text{Cell}} = E_{\text{Cathode}} - E_{\text{Anode}}$$

From the above, for the cell with the standard hydrogen electrode (potential of 0 by convention), one obtains:

$$E_{\text{Cell}} = E_{\text{Right}} - 0 = E_{\text{Electrode}}$$

The left-right convention is consistent with the international agreement that redox potentials be given for reactions written in the form of reduction half-reactions.

Electrode potential is measured in volts (V).

ORIGIN AND INTERPRETATION

Electrode potential appears at the interface between an electrode and electrolyte due to the transfer of charged species across the interface, specific adsorption of ions at the interface, and specific adsorption/orientation of polar molecules, including those of the solvent.

Electrode potential is the electric potential on an electrode component. In a cell, there will be an electrode potential for the cathode, and an electrode potential for the anode. The difference between the two electrode potentials equals the cell potential.

$$E_{\text{Cell}} = E_{\text{Cathode}} - E_{\text{Anode}}$$

The measured electrode potential may be either that at equilibrium on the working electrode ("reversible potential"), or a potential with a non-zero net reaction on the working electrode but zero net current ("corrosion potential", "mixed potential"), or a potential with a non-zero net current on the working electrode (like in galvanic corrosion or voltammetry). Reversible potentials can be sometimes converted to the standard electrode potential for a given electroactive species by extrapolation of the measured values to the standard state.

The value of the electrode potential under non-equilibrium depends on the nature and composition of the contacting phases, and on the kinetics of electrode reactions at the interface.

An operational assumption for determinations of the electrode potentials with the standard hydrogen electrode involves this reference electrode with hydrogen ion in an ideal solution having is "zero potential at all temperatures" equivalently to standard enthalpy of formation of hydrogen ion is also "zero at all temperatures".

MEASUREMENT

The measurement is generally conducted using a three-electrode setup:

- Working electrode
- Counter electrode
- Reference electrode (standard hydrogen electrode or an equivalent)

Working Electrode

The working electrode is the electrode in an electrochemical system on which the reaction of interest is occurring. The working electrode is often used in conjunction with an auxiliary electrode, and a reference electrode in a three electrode system. Depending on whether the reaction on the electrode is a reduction or an oxidation, the working electrode is called cathodic or anodic, respectively. Common working electrodes can consist of materials ranging from inert metals such as gold, silver or platinum, to inert carbon such as glassy carbon or pyrolytic carbon, and mercury drop and film electrodes. Chemically modified electrodes are employed for the analysis of both organic and inorganic samples.

Special Types of Working Electrodes

- Ultramicroelectrode (UME)
- Rotating disk electrode (RDE)
- Rotating ring-disk electrode (RRDE)
- Hanging mercury drop electrode (HMDE)
- Dropping mercury electrode (DME)

Ultramicroelectrode

An ultramicroelectrode (UME) is a working electrode used in a voltammetry. The small size of UME give them large diffusion layers and small overall currents. These features allow UME to achieve useful steady-state conditions and very high scan rates (V/s) with limited distortion. UME were developed independently by Wightman and Fleischmann around 1980. Small current at UME enables electrochemical measurements in low conductive media (organic solvents), where voltage drop associated with high solution resistance makes these experiments difficult for conventional electrodes. Furthermore, small voltage drop at UME leads to a very small voltage distortion at the electrode-solution interface which allows using two-electrode setup in voltammetric experiment instead of conventional three-electrode setup.

Design

Ultramicroelectrodes are often defined as electrodes which are smaller than the diffusion layer achieved in a readily accessed experiment. A working definition is an electrode that has at least one dimension (the critical dimension) smaller than 25 μm. Platinum electrodes with a radius of 5 μm are commercially available and electrodes with critical dimension of 0.1 μm have been made. Electrodes with even smaller critical dimension have been reported in the literature, but exist mostly as proofs of concept. The most common UME is a disk shaped electrode created by embedding a thin wire in glass, resin, or plastic. The resin is cut and polished to expose a cross section of the wire. Other shapes, such as wires and rectangles, have also been reported. Carbon-fiber microelectrodes are fabricated with conductive carbon carbon fibers sealed in glass capillary with exposed tips. These electrodes are frequently used with in vivo voltammetry.

Theory

Linear Region

Every electrode has a range of scan rates called the linear region. The response to a reversible redox couple in the linear region is a "diffusion controlled peak" which can be modeled with the Cottrell equation. The upper limit of the useful linear region is bound by an excess of charging current combined with distortions created from large peak currents and associated resistance. The charging current scales linearly with scan rate while the peak current, which contains the useful information, scales with the square root of scan rate. As scan rates increase, the relative peak response diminishes. Some of the charge current can be mitigated with RC compensation and/or mathematically removed after the experiment. However, the distortions resulting from increased current and the associated resistance cannot be subtracted. These distortions ultimately limit the scan rate for which an electrode is useful. For example, a working electrode with a radius of 1.0 mm is not useful for experiments much greater than 500 mV/s.

Moving to an UME drops the currents being passed and thus greatly increases the useful sweep rate up to 10^6 V/s. These faster scan rates allow the investigation of electrochemical reaction mechanisms with much higher rates than can be explored with regular working electrodes. By adjusting the size of the working electrode an enormous kinetic range can be studied. For UME only the very fast reactions can be studied through peak current since the linear region only exists for UME at very high scan rates.

Steady-State Region

At scan rates slower than those of the linear region is a region which is mathematically complex to model and rarely investigated. At even slower scan rates there is the steady-state region. In the steady-state region linear sweeps traces display reversible redox couple as steps rather than peaks. These steps can readily be modeled for meaningful data.

To access the steady-state region the scan rate must be dropped. As scan rates are slowed, the relative currents also drop at a given point reducing the reliability of the measurement. The low ratio of diffusion layer volume to electrode surface area means regular stationary electrodes cannot be dropped low enough before their current measurements become unreliable. In contrast, the ratio of diffusion layer volume to electrode surface area is much higher for UME. When the scan rate of UME is dropped it quickly enters the steady-state regime at useful scan rates. Even though UME supply small total currents their steady-state currents are high compared to regular electrodes.

Rotating Disk Electrode

A rotating disk electrode (RDE) is a hydrodynamic working electrode used in a three electrode system. The electrode rotates during experiments inducing a flux of analyte to the electrode. These working electrodes are used in electrochemical studies when investigating reaction mechanisms related to redox chemistry, among other chemical phenomena. The more complex rotating ring-disk electrode can be used as a *rotating disk electrode* if the ring is left inactive during the experiment.

Structure

The electrode includes a conductive disk embedded in an inert non-conductive polymer or resin that can be attached to an electric motor that has very fine control of the electrode's rotation rate. The disk, like any working electrode, is generally made of a noble metal or glassy carbon, however any conductive material can be used based on specific needs.

Function

The disk's rotation is usually described in terms of angular velocity. As the disk turns, some of the solution described as the *hydrodynamic boundary layer* is dragged by the spinning disk and the resulting centrifugal force flings the solution away from the center of the electrode. Solution flows up, perpendicular to the electrode, from the bulk to replace the *boundary layer*. The sum result is a laminar flow of solution towards and across the electrode. The rate of the solution flow can be controlled by the electrode's angular velocity and modeled mathematically. This flow can quickly achieve conditions in which the steady-state current is controlled by the solution flow rather than diffusion. This is a contrast to still and unstirred experiments such as cyclic voltammetry where the steady-state current is limited by the diffusion of species in solution.

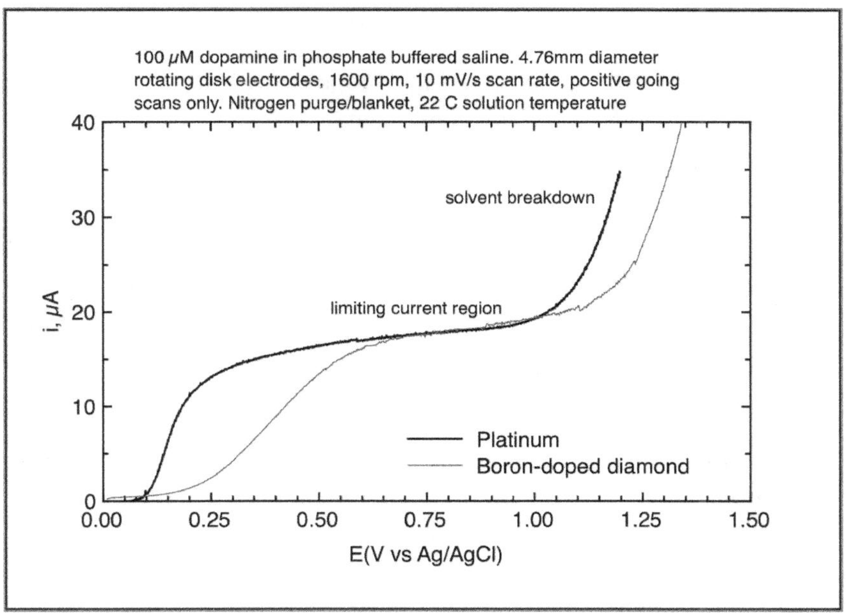

Linear sweep voltammogram of 100 μM dopamine in phosphate buffered saline. Adapted from D. Sopchak, PhD thesis, Case Western Reserve University, 2001, figure 4.1

By running linear sweep voltammetry and other experiments at various rotation rates, different electrochemical phenomena can be investigated, including multi-electron transfer, the kinetics of a slow electron transfer, adsorption/desorption steps, and electrochemical reaction mechanisms.

Differences in Behavior From Stationary Electrodes

Potential sweep reversals as used in cyclic voltammetry are different for a RDE system, since the products of the potential sweep are continually swept away from the electrode. A reversal would produce a similar *i-E* curve, which would closely match the forward scan, except for capacitive charging current. A RDE cannot be used to observe the behavior of the electrode reaction products, since they are continually swept away from the electrode. However, the rotating ring-disk electrode is well suited to investigate this further reactivity. The peak current in a cyclic voltammogram for a RDE is a plateau like region, governed by the Levich Equation. The limiting current is typically much higher than the peak current of a stationary electrode, being that the mass transport of reactants is actively stimulated by the rotating disk, and not just governed by diffusion, as is the case for a stationary electrode. Any rotating disk electrode can, of course, also be used as a stationary electrode by using it with the rotator turned off.

Rotating Ring-Disk Electrode

A rotating ring-disk electrode (RRDE) is double working electrode used in hydrodynamic voltammetry, very similar to a rotating disk electrode (RDE). The electrode actually rotates during experiments inducing a flux of analyte to the electrode. These working electrode are used in electrochemical studies when investigating reaction mechanisms related to redox chemistry among other chemical phenomena.

Structure

The difference between a rotating ring-disk electrode and a rotating disk electrode is the addition of a second working electrode in the form of a ring around the central disk of the first working electrode. The two electrodes are separated by a non-conductive barrier and connected to the potentiostat through different leads. To operate such an electrode it is necessary to use a bipotentiostat or some potentiostat capable of controlling a four electrode system. This rotating hydrodynamic electrode motif can be extend to *rotating double-ring electrodes* and *rotating double-ring-disk electrodes* and other even more esoteric constructions as suited to a given experiment.

Function

The RRDE takes advantage of the form of the laminar flow created during rotation. As the system is rotated the solution in contact with the electrode is driven to the side of the electrode the same as with a rotating disk electrode. As the solution flows to the side it crosses the ring electrode and back into the bulk of the solution. If the flow in the solution is laminar then the solution is brought in contact with the disk quickly followed by the ring in a very controlled manner. The resulting currents are dependent on the electrodes' respective potentials, areas, and spacing as well as the rotation rate and given substrate.

This design makes a variety of experiments possible, for example a complex could be oxidized at the disk and then reduced back to the starting material at the ring. It is easy to predict what the ring/disk current ratios is if this process is entirely controlled by the flow of solution. If it is not controlled by the flow of the solution the current will deviate. For example, if the first oxidation is followed by a chemical reaction, an EC mechanism, to form a product that can not be reduced at the ring then the magnitude of the ring current would be reduced. By varying the rate of rotation it is possible to determine the rate of the chemical reaction if its in the proper kinetic regime.

Applications

The RRDE setup allows for many additional experiments well beyond the capacity of a RDE. For example, while one electrode conducts linear sweep voltammetry the other can be kept at a constant potential or also swept in a controlled manner. Step experiments with each electrode acting independently can be conducted. These as well as many other extremely elegant experiments are possible, including those tailored to the needs of a given system. Such experiments are useful in studying multi-electrons processes, the kinetics of a slow electron transfer, adsorption/desorption steps, and electrochemical reaction mechanisms.

The RRDE is an important tool for characterizing the fundamental properties of electrocatalysts used in fuel cells. For example, in a proton exchange membrane (PEM) fuel cell, dioxygen reduction at the cathode is often enhanced by an electrocatalyst comprising platinum nanoparticles. When oxygen is reduced using an electrocatalyst, an unwanted and harmful by-product, hydrogen peroxide, may be produced. Hydrogen peroxide can damage the internal components of a PEM fuel cell, so oxygen-reduction electrocatalysts are engineered in such a way as to limit the amount of peroxide formed. An RRDE "collection experiment" can be used to probe the peroxide generating tendencies of an electrocatalyst. In this experiment, the disk is coated with a thin layer bearing the electrocatalyst, and the disk electrode is poised at a potential which reduces the oxygen. Any products generated at the disk electrode are then swept past the ring electrode. The potential of the ring electrode is poised to detect any hydrogen peroxide that may have been generated at the disk.

Design Considerations

In general, narrowing the gap between the disk outer diameter and the ring inner diameter allows probing of systems with faster kinetics. A narrow gap reduces the "transit time" necessary for an intermediate species generated at the disk to successfully reach the ring electrode and be detected. Using precision machining techniques, it is possible to make gaps between 0.1 and 0.5 millimeters, and narrower gaps have been created using microlithography techniques.

Another important parameter for an RRDE is the "collection efficiency". This parameter is a measure of the percentage of the material generated at the disk electrode which is detected at the ring electrode. For any given set of RRDE dimensions (disk OD, ring ID, and ring OD), the collection efficiency can be computed using formulae derived from fluid dynamics first principles. One useful aspect of the theoretical collection efficiency is that it is only a function of the RRDE dimensions. That is, it is independent of the rotation rate over a wide range of rotation rates.

It is desirable for an RRDE to have a large collection efficiency if only to assure that the current signal measured at the ring electrode is detectable. On the other hand, it also desirable for an RRDE to have a small transit time so that short-lived (unstable) intermediate products generated at the disk survive long enough to be detected at the ring. The choice of actual RRDE dimensions is often a trade-off between a large collection efficiency or a short transit time.

Hanging Mercury Drop Electrode

The hanging mercury drop electrode (HMDE or HDME) is a working electrode variation on the dropping mercury electrode (DME). Experiments run with dropping mercury electrodes are referred to as forms of polarography. If the experiments are performed at an electrode with a constant surface (like the HMDE) it is referred as voltammetry.

Like other working electrodes these electrodes are used in electrochemical studies using three electrode systems when investigating reaction mechanisms related to redox chemistry among other chemical phenomenon.

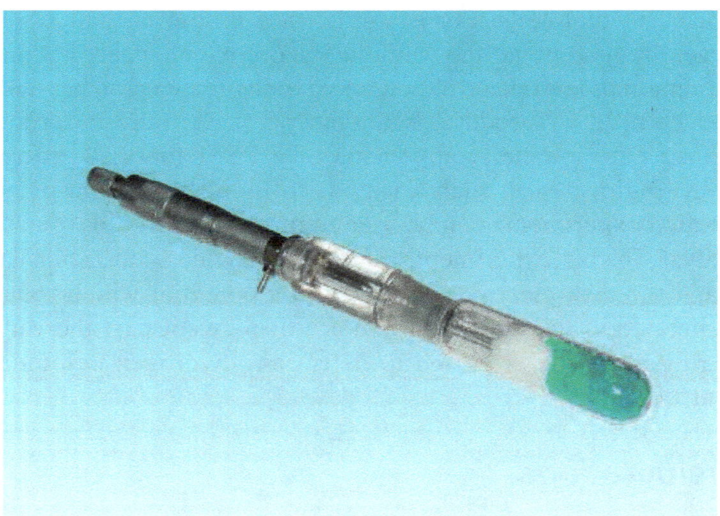

Hanging Mercury Drop Electrode.

Distinction

The hanging mercury drop electrode (HMDE) produces a partial mercury drop of controlled geometry and surface area at the end of a capillary in contrast to the dropping mercury electrode (DME) which steadily releases drops of mercury during an experiment. The disadvantages a DME experiences due to a constantly changing surface are not experienced by the HMDE since it has static surface area during an experiment. The static surface of the HMDE means it is more likely to suffer from the surface adsorption phenomenon than a DME. Unlike solid electrodes which need to be cleaned and polished between most experiments, the self-renewing HMDE can simply release the contaminated drop and grow a clean drop between each experiment.

Dropping Mercury Electrode

The dropping mercury electrode (DME) is a working electrode made of mercury and used in polarography. Experiments run with mercury electrodes are referred to as forms of polarography even if the experiments are identical or very similar to a corresponding voltammetry experiment which uses solid working electrodes. Like other working electrodes these electrodes are used in electrochemical studies using three electrode systems when investigating reaction mechanisms related to redox chemistry among other chemical phenomena.

DME.

Structure

A flow of mercury passes through an insulating capillary producing a droplet which grows from the end of the capillary in a reproducible way. Each droplet grows until it reaches a diameter of about a millimeter and releases. The released droplet is no longer in contact with the working electrode whose contact is above the capillary. As the electrode is used mercury collects in the bottom of the cell. In some cell designs this mercury pool is connected to a lead and used as the cell's auxiliary electrode. Each released drop is immediately followed by the formation of another drop. The drops are generally produced at a rate of about 0.2 Hz.

Considerations

A major advantage of the DME is that each drop has a smooth and uncontaminated surface free from any adsorbed analyte or impurity. The self-renewing electrode does not need to be cleaned or polished like a solid electrode. This advantage comes at the cost of a working electrode with a constantly changing surface area. Since the drops are produced predictably the changing surface area can be accounted for or even used advantageously. In addition the drops' growth causes capacitive current and faradaic current. These changing current effects combined with experiments where the potential is continuously changed can result in noisy traces. In some experiments the traces are continually sampled, showing all the current deviation resulting from the drop growth. Other sampling methods smooth the data by sampling the current at the electrode only once per drop at a specific size. The DME's periodic expansion into the solution and hemispherical shape also affects the way the analyte diffuses to the electrode surface. The DME consists of a fine capillary with a bore size of 20-50 μm.

Auxiliary Electrode

The auxiliary electrode, often also called the counter electrode, is an electrode used in a three electrode electrochemical cell for voltammetric analysis or other reactions in which an electric current is expected to flow. The auxiliary electrode is distinct from the reference electrode, which establishes the electrical potential against which other potentials may be measured, and the working electrode, at which the cell reaction takes place.

In a two-electrode system, either a known current or potential is applied between the working and auxiliary electrodes and the other variable may be measured. The auxiliary electrode functions as a cathode whenever the working electrode is operating as an anode and vice versa. The auxiliary electrode often has a surface area much larger than that of the working electrode to ensure that the half-reaction occurring at the auxiliary electrode can occur fast enough so as not to limit the process at the working electrode.

When a three electrode cell is used to perform electroanalytical chemistry, the auxiliary electrode, along with the working electrode, provides circuit over which current is either applied or measured. Here, the potential of the auxiliary electrode is usually not measured and is adjusted so as to balance the reaction occurring at the working electrode. This configuration allows the potential of the working electrode to be measured against a known reference electrode without compromising the stability of that reference electrode by passing current over it.

The auxiliary electrode may be isolated from the working electrode using a glass frit. Such isolation prevents any byproducts generated at the auxiliary electrode from contaminating the main test solution: for example, if a reduction is being performed at the working electrode in aqueous solution, oxygen may be evolved from the auxiliary electrode. Such isolation is crucial during the bulk electrolysis of a species which exhibits reversible redox behavior.

Auxiliary electrodes are often fabricated from electrochemically inert materials such as gold, platinum, or carbon.

Reference Electrode

A reference electrode is an electrode which has a stable and well-known electrode potential. The high stability of the electrode potential is usually reached by employing a redox system with constant (buffered or saturated) concentrations of each participants of the redox reaction.

There are many ways reference electrodes are used. The simplest is when the reference electrode is used as a half cell to build an electrochemical cell. This allows the potential of the other half cell to be determined. An accurate and practical method to measure an electrode's potential in isolation (absolute electrode potential) has yet to be developed.

Aqueous Reference Electrodes

Common reference electrodes and potential with respect to the standard hydrogen electrode:

- Standard hydrogen electrode (SHE) (E=0.000 V) activity of H^+=1
- Normal hydrogen electrode (NHE) (E ≈ 0.000 V)concentration H^+=1
- Reversible hydrogen electrode (RHE) (E=0.000 V - 0.0591*pH)
- Saturated calomel electrode (SCE) (E=+0.241 V saturated)
- Copper-copper(II) sulfate electrode(CSE) (E=+0.314 V)
- Silver chloride electrode (E=+0.197 V saturated)
- pH-electrode (in case of pH buffered solutions)
- Palladium-hydrogen electrode
- Dynamic hydrogen electrode (DHE)
- Mercury-mercurous sulfate electrode (E=+0.64 V in sat'd K_2SO_4, E=+0.68 V in 0.5 M H_2SO_4) (MSE)

Standard Hydrogen Electrode

The Standard hydrogen electrode (abbreviated SHE), is a redox electrode which forms the basis of the thermodynamic scale of oxidation-reduction potentials. Its absolute electrode potential is estimated to be 4.44 ± 0.02 V at 25 °C, but to form a basis for comparison with all other electrode reactions, hydrogen›s standard electrode potential (E^0) is declared to be zero volts at all temperatures. Potentials of any other electrodes are compared with that of the standard hydrogen electrode at the same temperature.

Hydrogen electrode is based on the redox half cell:

$2\ H^+(aq) + 2\ e^- \rightarrow H_2(g)$

This redox reaction occurs at a platinized platinum electrode. The electrode is dipped in an acidic solution and pure hydrogen gas is bubbled through it. The concentration of both the reduced form and oxidised form is maintained at unity. That implies that the pressure of hydrogen gas is 1 bar (100 kPa) and the activity of hydrogen ions in the solution is unity. The activity of hydrogen ions is their effective concentration, which is equal to the formal concentration times the activity coefficient. These unit-less activity coefficients are close to 1.00 for very dilute water solutions, but usually lower for more concentrated solutions. The Nernst equation should be written as:

$$E = \frac{RT}{F} \ln \frac{a_{H^+}}{\sqrt{p_{H_2}/p^0}}$$

$$E = -\frac{2.303RT}{F}\text{pH} - \frac{RT}{2F} \ln \frac{p_{H_2}}{p^0}$$

where:

- a_H^+ is the activity of the hydrogen ions, $a_H^+ = f_H^+C_H^+ / C^0$
- p_{H2} is the partial pressure of the hydrogen gas, in pascals, Pa
- R is the universal gas constant
- T is the temperature, in kelvins
- F is the Faraday constant (the charge per mole of electrons), equal to 9.6485309 $\times 10^4$ C mol^{-1}
- p^0 is the standard pressure, 10^5 Pa

Relationship between the normal hydrogen electrode (NHE) and the standard hydrogen electrode (SHE)

During the early development of electrochemistry, researchers used the normal hydrogen electrode as their standard for zero potential. This was convenient because it could *actually be constructed* by having "[immersing] a platinum electrode into a solution of 1 N strong acid and [bubbling] hydrogen gas through the solution at about 1 atm pressure". However, this electrode/solution interface was later changed. What replaced it was a theoretical electrode/solution interface, where the concentration of H^+ was 1 M, but the H^+ ions were assumed to have no interaction with other ions (a condition not physically attainable at those concentrations). To differentiate this new standard from the previous one it was given the name 'Standard Hydrogen Electrode'.

In summary,

NHE: potential of a platinum electrode in 1 N acid solution

SHE: potential of a platinum electrode in a theoretical ideal solution (the current standard for zero potential for all temperatures)

Choice of Platinum

The choice of platinum for the hydrogen electrode is due to several factors:

- inertness of platinum (it does not corrode)
- the capability of platinum to catalyze the reaction of proton reduction
- a high intrinsic exchange current density for proton reduction on platinum
- excellent reproducibility of the potential (bias of less than 10 μV when two well-made hydrogen electrodes are compared with one another)

The surface of platinum is platinized (i.e., covered with platinum black) to:

- Increase total surface area. This improves reaction kinetics and maximum possible current
- Use a surface material that adsorbs hydrogen well at its interface. This also improves reaction kinetics

Other metals can be used for building electrodes with a similar function such as the palladium-hydrogen electrode.

Interference

Because of the high adsorption activity of the platinized platinum electrode, it's very important to protect electrode surface and solution from the presence of organic substances as well as from atmospheric oxygen. Inorganic ions that can reduce to a lower valency state at the electrode also have to be avoided (e.g., Fe^{3+}, CrO2−4). A number of organic substances are also reduced by hydrogen at a platinum surface, and these also have to be avoided. Cations that can reduce and deposit on the platinum can be source of interference: silver, mercury, copper, lead, cadmium and thallium. Substances that can inactivate ("poison") the catalytic sites include arsenic, sulfides and other sulfur compounds, colloidal substances, alkaloids, and material found in living systems.

Isotopic Effect

The standard redox potential of the deuterium couple is slightly different from that of the proton couple (ca. −0.0044 V vs SHE). Various values in this range have been obtained: −0.0061 V, −0.00431 V, −0.0074 V.

$2\ D^+(aq) + 2\ e^- \rightarrow D_2(g)$ Also difference occur when hydrogen deuteride is used instead of hydrogen in the electrode.

Construction

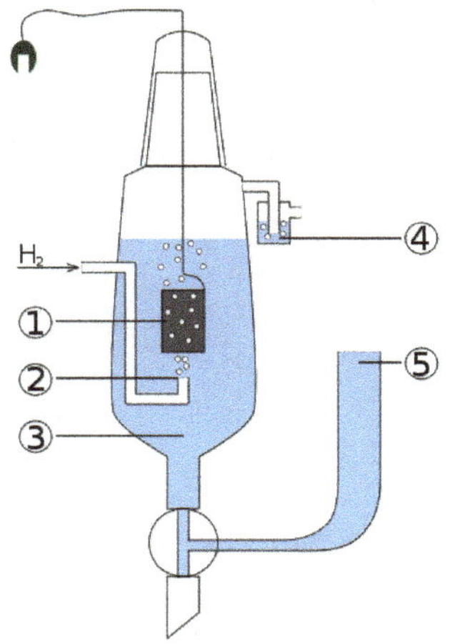

The scheme of the standard hydrogen electrode:

- platinized platinum electrode
- hydrogen blow
- solution of the acid with activity of $H^+ = 1$ mol dm^{-3}
- hydroseal for prevention of the oxygen interference
- reservoir through which the second half-element of the galvanic cell should be attached. The connection can be direct, through a narrow tube to reduce mixing, or through a salt bridge, depending on the other electrode and solution. This creates an ionically conductive path to the working electrode of interest.

Reversible Hydrogen Electrode

A reversible hydrogen electrode (RHE) is a reference electrode, more specifically a subtype of the standard hydrogen electrodes, for electrochemical processes. Unlike the standard hydrogen electrode, its measured potential does not change with the pH, so it can be directly used in the electrolyte.

The name refers to the fact that the electrode is in the actual electrolyte solution and not separated by a salt bridge. The hydrogen ion concentration is therefore not 1, but corresponds to that of the electrolyte solution; in this way we can achieve a stable potential with a changing pH value. The potential of the RHE correlates to the pH value:

$$E_0 = 0.000 - 0.059 \times \mathrm{pH}$$

In general, for hydrogen electrodes in which the reaction:

$$2\,H_3O^+ + 2\,e^- \rightleftharpoons H_2 + 2\,H_2O$$

expires, the following dependence of the equilibrium potential E, hydrogen pressure $p[H_2]$ and the activity $a[H_3O^+]$of the hydronium ions:

$$E = E_{00} + \frac{RT}{F}\left(\ln(a[H_3O^+]) - \tfrac{1}{2}\ln(p[H_2])\right)$$

Here E_{00} is the standard reduction potential (this is by definition equal to zero), R is the universal gas constant, T the absolute temperature and F is the Faraday constant.

Surges occur in the electrolysis of water which means that the required cell voltage due to kinetic inhibition is higher than the equilibrium potential. The voltage increases with increasing current density at the electrodes. The measurement of equilibrium potentials is therefore possible without power.

Principle

The reversible hydrogen electrode is a fairly practical and reproducible electrode "standard." The term refers to a hydrogen electrode immersed in the electrolyte solution actually used.

The benefit of that electrode is that no salt bridge is needed:

- no contamination of the electrolyte by Cl^- or SO_4^{2-}
- no diffusion potentials at the electrolyte bridge (liquid junction potential). This is important at temperature different to 25 °C.
- long time measurements possible (no electrolyte bridge means no maintenance of the bridge).

Saturated Calomel Electrode

The Saturated calomel electrode (SCE) is a reference electrode based on the reaction between elemental mercury and mercury(I) chloride. It has been widely replaced by the silver chloride electrode, however the calomel electrode has a reputation of being more robust. The aqueous phase in contact with the mercury and the mercury(I) chloride (Hg_2Cl_2, "calomel") is a saturated solution of potassium chloride in water. The electrode is normally linked *via* a porous frit to the solution in which the other electrode is immersed. This porous frit is a salt bridge.

In cell notation the electrode is written as:

$Cl^-(4\,M)|Hg_2Cl_2(s)|Hg(l)|Pt$

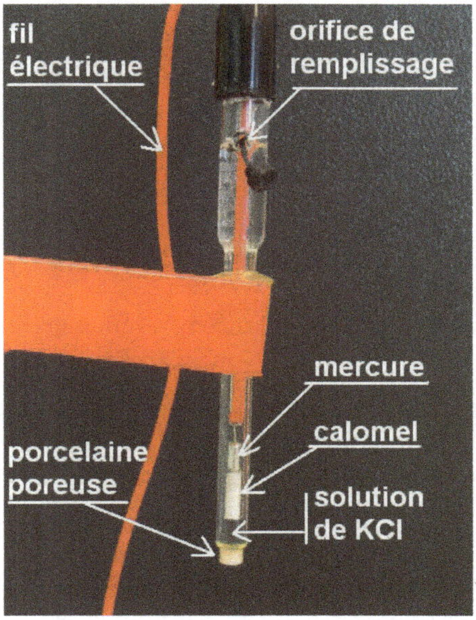

Saturated calomel electrode.

Theory of Operation

The electrode is based on the redox reaction

$$Hg_2^{2+} + 2\,e^- \rightleftharpoons 2\,Hg(l)$$

The Nernst equation for this reaction is

$$E = E^0_{Hg_2^{2+}/Hg} - \frac{RT}{2F}\ln\frac{1}{a_{Hg_2^{2+}}}$$

where E^0 is the standard electrode potential for the reaction and a_{Hg} is the activity for the mercury cation (the activity for a liquid of 1 Molar is 1). This activity can be found from the solubility product of the reaction

$$Hg_2^{2+} + 2\,Cl^- \rightleftharpoons Hg_2Cl_2(s), \qquad K_{sp} = a_{Hg_2^{2+}}\,a^2_{Cl^-}$$

By replacing the activity in the Nernst equation with the value in the solubility equation, we get

$$E = E^0_{Hg_2^{2+}/Hg} + \frac{RT}{2F}\ln K_{sp} - \frac{RT}{2F}\ln a^2_{Cl^-}$$

The only variable in this equation is the activity (or concentration) of the chloride anion. But since the inner solution is saturated with potassium chloride, this activity is fixed by the solubility of potassium chloride. When saturated the redox potential of the calomel electrode is +0.2444 V vs. SHE at 25 °C, but slightly higher when the chloride solution is less than saturated. For example, a 3.5M KCl electrolyte solution increases the reference potential to +0.250 V vs. SHE at 25 °C, and a 0.1 M solution to +0.3356 V at the same temperature.

Application

The SCE is used in pH measurement, cyclic voltammetry and general aqueous electrochemistry.

This electrode and the silver/silver chloride reference electrode work in the same way. In both electrodes, the activity of the metal ion is fixed by the solubility of the metal salt.

The calomel electrode contains mercury, which poses much greater health hazards than the silver metal used in the Ag/AgCl electrode.

Copper–Copper(II) Sulfate Electrode

The copper–copper(II) sulfate electrode is a reference electrode of the first kind, based on the redox reaction with participation of the metal (copper) and its salt, copper(II) sulfate. It is used for measuring electrode potential and is the most commonly used reference electrode for testing cathodic protection corrosion control systems. The corresponding equation can be presented as follow:

$Cu^{2+} + 2e^{-} \rightarrow Cu^{0}$(metal)

This reaction characterized by reversible and fast electrode kinetics, meaning that a sufficiently high current can be passed through the electrode with the 100% efficiency of the redox reaction (dissolution of the metal or cathodic deposition of the copper-ions).

The Nernst equation below shows the dependence of the potential of the copper-copper(II) sulfate electrode on the activity or concentration copper-ions:

$$E = 0.337 + \frac{RT}{2F} \ln a_{Cu^{2+}}$$

Commercial reference electrodes consist of a plastic tube holding the copper rod and saturated solution of copper sulfate. A porous plug on one end allows contact with the copper sulfate electrolyte. The copper rod protrudes out of the tube. A voltmeter negative lead is connected to the copper rod.

The potential of a copper copper sulfate electrode is +0.314 volt with respect to the standard hydrogen electrode. Copper–copper(II) sulfate electrode is also used as one of the half cells in the Daniel-Jakobi galvanic cell.

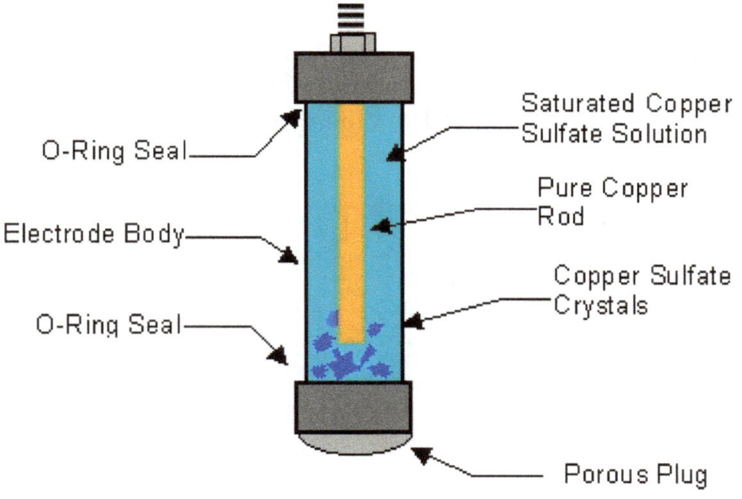

Cu-Cu(II) reference electrode.

Silver Chloride Electrode

A silver chloride electrode is a type of reference electrode, commonly used in electrochemical measurements. For environmental reasons it has widely replaced the saturated calomel electrode. For example, it is usually the internal reference electrode in pH meters and it is often used as reference in reduction potential measurements. As an example of the latter, the silver chloride electrode is the most commonly used reference electrode for testing cathodic protection corrosion control systems in sea water environments.

The electrode functions as a redox electrode and the equilibrium is between the silver metal (Ag) and its salt — silver chloride (AgCl, also called silver(I) chloride).

The corresponding equations can be presented as follows:

$$Ag^+ + e^- \rightleftharpoons Ag(s)$$
$$AgCl(s) \rightleftharpoons Ag^+ + Cl^-$$

or an overall reaction can be written:

$$AgCl(s) + e^- \rightleftharpoons Ag(s) + Cl^-$$

This reaction is characterized by fast electrode kinetics, meaning that a sufficiently high current can be passed through the electrode with the 100% efficiency of the redox reaction (dissolution of the metal or cathodic deposition of the silver-ions). The reaction has been proven to obey these equations in solutions of pH values between 0 and 13.5.

The Nernst equation below shows the dependence of the potential of the silver-silver(I) chloride electrode on the activity or effective concentration of chloride-ions:

$$E = E^0 - \frac{RT}{F} \ln a_{Cl^-}$$

The standard electrode potential E^0 against standard hydrogen electrode (SHE) is 0.230V ± 10mV. The potential is however very sensitive to traces of bromide ions which make it more negative. (The more exact standard potential given by an IUPAC review paper is +0.22249 V, with a standard deviation of 0.13 mV at 25 °C.).

Ag-AgCl reference electrode.

Applications

Commercial reference electrodes consist of a plastic tube electrode body. The electrode is a silver wire that is coated with a thin layer of silver chloride, either physically by dipping the wire in molten silver chloride, chemically by electroplating the wire in concentrated hydrochloric acid or electrochemically by oxidizing the silver in a chloride solution.

A porous plug on one end allows contact between the field environment with the silver chloride electrolyte. An insulated lead wire connects the silver rod with measuring instruments. A voltmeter negative lead is connected to the test wire.

The electrode body contains potassium chloride to stabilize the silver chloride concentration. When working in seawater, this body can be removed and the chloride concentration is fixed by the stable salinity of the water. The potential of a silver:silver chloride reference electrode with respect to the standard hydrogen electrode depends on the electrolyte composition.

Reference Electrode Potentials		
Electrode	**Potential E^0+E_{lj}**	**Temperature Coef.**
	(V) at 25 °C	(mV/°C) at around 25 °C
SHE	0.000	0.000
Ag/AgCl/Sat. KCl	+0.197	-1.01
Ag/AgCl/3.5 mol/kg KCl	+0.205	-0.73
Ag/AgCl/3.0 mol/kg KCl	+0.210	?
Ag/AgCl/1.0 mol/kg KCl	+0.235	+0.25
Ag/AgCl/0.6 mol/kg KCl	+0.25	
Ag/AgCl (Seawater)	+0.266	

Notes to the Table: (1) The table data source is, except where a separate reference is given. (2) E_{lj} is the potential of the liquid junction between the given electrolyte and the electrolyte with the activity of chloride of 1 mol/kg.

The electrode has many features making is suitable for use in the field:

- Simple construction
- Inexpensive to manufacture
- Stable potential
- Non-toxic components

They are usually manufactured with saturated potassium chloride electrolyte, but can be used with lower concentrations such as 1 mol/kg potassium chloride. As noted above, changing the electrolyte concentration changes the electrode potential. Silver chloride is slightly soluble in strong potassium chloride solutions, so it is sometimes recommended the potassium chloride be saturated with silver chloride to avoid stripping the silver chloride off the silver wire.

Biological Electrode Systems

Silver chloride electrodes are also used by many applications of biological electrode systems such as biomonitoring sensors as part of electrocardiography (ECG) and electroencephalography (EEG), and in transcutaneous electrical nerve stimulation (TENS) to deliver current. Historically, the electrodes were fabricated from solid materials such as silver, brass coated with silver, tin and nickel. In today's applications, most biomonitoring electrodes are silver/silver chloride sensors which are fabricated by coating a thin layer of silver on plastic substrates and the outer layer of silver is converted to silver chloride.

Tab electrode using silver/silver chloride sensing for electrocardiography (ECG)

The principle of silver/silver chloride sensors operation is the conversion of ion current at the surface of human tissues to electron current to be delivered through the lead wire to the instrument to read. An important part of the operation is electrolyte gel which is applied between the electrode and tissues. The gel contains free chloride ions such that the charge can be carried through the electrolyte, therefore the electrolyte can be considered as conductive for ion current as the human tissues. When the ion current exists, the silver atoms in the electrode oxidize and discharge cations to the electrolyte and the electrons carry charge through the lead wire. At the same time, the chloride ions which are anions in the electrolyte travel toward the electrode and they are reduced as they bond with silver of the electrode resulting in silver chloride and free electrons to deliver to the lead wire. The reaction allows current to pass from electrolyte to electrode and the electron current passes through the lead wire for the instrument to read.

When there is an uneven distribution of cations and anions, there will be a small voltage called *half-cell potential* associated with the current. In the DC system that is used by the ECG and EEG instruments, the difference between the half-cell potential and the zero potential is shown as DC offset which is an undesirable characteristic. Silver/silver chloride is a popular choice of biological electrodes due to its low half-cell potential of approximately 220 mV and the lower of the impedance of the electrode by silver chloride.

Elevated Temperature Application

When appropriately constructed, the silver chloride electrode can be used up to 300 °C. The standard potential (i.e., the potential when the chloride activity is 1 mol/kg) of the silver chloride electrode is a function of temperature as follows:

Temperature Dependence of the Standard Potential of the Silver/Silver Chloride Electrode	
Temperature	**Potential E⁰**
°C	V versus SHE at the same temperature
25	0.22233
60	0.1968
125	0.1330
150	0.1032
175	0.0708
200	0.0348
225	-0.0051
250	-0.054
275	-0.090

Bard et al. give the following correlations for the standard potential of the silver chloride electrode as a function of temperature (where t is temperature in °C):

$E^0(V) = 0.23695 - 4.8564\text{x}10^{-4}t - 3.4205\text{x}10^{-6}t^2 - 5.869 \text{ x } 10^{-9}t^3$ for $0 < t < 95$ °C.

The same source also gives the fit to the high-temperature potential, which reproduces the data in the table above:

$E^0(V) = 0.23735 - 5.3783\text{x}10^{-4}t - 2.3728\text{x}10^{-6}t^2$ for $25 < t < 275$ °C.

The extrapolation to 300 °C gives E^0 of -0.138 V.

Farmer gives the following correlation for the potential of the silver chloride electrode with 0.1 mol/kg KCl solution, accounting for the activity of Cl^- at the elevated temperature:

$E^{0.1 \text{ mol/kg KCl}}(V) = 0.23735 - 5.3783\text{x}10^{-4}t - 2.3728\text{x}10^{-6}t^2 + 2.2671\text{x}10^{-4}(t+273)$ for $25 < t < 275$ °C.

Glass Electrode

A glass electrode is a type of ion-selective electrode made of a doped glass membrane that is sensitive to a specific ion. The most common application of ion-selective glass electrodes is for the measurement of pH. The pH electrode is an example of a glass electrode that is sensitive to hydrogen ions. Glass electrodes play an important part in the instrumentation for chemical analysis and physico-chemical studies. The voltage of the glass electrode, relative to some reference value, is sensitive to changes in the activity of certain type of ions.

History

The first studies of glass electrodes (GE) found different sensitivities of different glasses to change of the medium's acidity (pH), due to effects of the alkali metal ions.

- 1906 — M. Cremer determined that the electric potential that arises between parts of the fluid, located on opposite sides of the glass membrane is proportional to the concentration of acid (hydrogen ion concentration).
- 1909 — S. P. L. Sørensen introduced the concept of pH.
- 1909 — F. Haber and Z. Klemensiewicz publicized on 28 January 1909 results of their research on the glass electrode in The Society of Chemistry in Karlsruhe (first publication — The Journal of Physical Chemistry by W. Ostwald and J. H. van 't Hoff) — 1909).
- 1922 — W. S. Hughes showed that the alkali-silicate GE are similar to hydrogen electrode, reversible with respect to H^+.
- 1925 — P.M. Tookey Kerridge developed the first glass electrode for analysis of blood samples, and highlighted some of the practical problems with the equipment such as the high resistance of glass (50-150 megaOhms). During her PhD, Kerridge developed the miniature glass electrode, maximising the surface area of the tool by heat treating the platinum with platinum chloride at red heat, thus enabling a much larger signal: her design was the predecessor of many of the glass electrodes used today.

Applications

Glass electrodes are commonly used for pH measurements. There are also specialized ion sensitive glass electrodes used for determination of concentration of lithium, sodium, ammonium, and other ions. Glass electrodes have been utilized in a wide range of applications — from pure research, control of industrial processes, to analyze foods, cosmetics and comparison of indicators of the environment and environmental regulations: a microelectrode measurements of membrane electrical potential of a biological cell, analysis of soil acidity, etc.

Types

Almost all commercial electrodes respond to single charged ions, like H^+, Na^+, Ag^+. The most common glass electrode is the pH-electrode. Only a few chalcogenide glass electrodes are sensitive to double-charged ions, like Pb^{2+}, Cd^{2+} and some others.

There are two main glass-forming systems:

- silicate matrix based on molecular network of silicon dioxide (SiO_2) with additions of other metal oxides, such as Na, K, Li, Al, B, Ca, etc.
- chalcogenide matrix based on molecular network of AsS, AsSe, AsTe.

Interfering Ions

Because of the ion-exchange nature of the glass membrane, it is possible for some other ions to concurrently interact with ion-exchange centers of the glass and to distort the linear dependence of the measured electrode potential on pH or other electrode function. In some cases it is possible to change the electrode function from one ion to another. For example, some silicate pNa electrodes can be changed to pAg function by soaking in a silver salt solution.

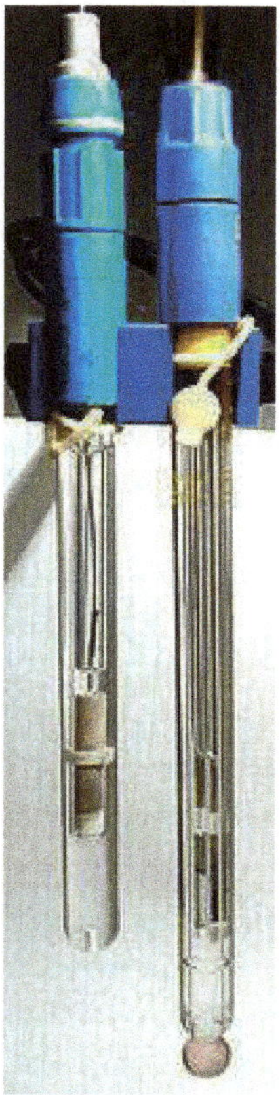

A silver chloride reference electrode (left) and glass pH electrode (right)

Interference effects are commonly described by the semiempirical Nicolsky-Eisenman equation (also known as Nikolsky-Eisenman equation), an extension to the Nernst equation. It is given by

$$E = E^0 + \frac{RT}{z_i F} \ln \left[a_i + \sum_j \left(k_{ij} a_j^{z_i/z_j} \right) \right]$$

where E is the emf, E^0 the standard electrode potential, z the ionic valency including the sign, a the activity, i the ion of interest, j the interfering ions and k_{ij} is the selectivity coefficient. The smaller the selectivity coefficient, the less is the interference by j.

To see the interfering effect of Na^+ to a pH-electrode:

$$E = E^0 + \frac{RT}{F}\ln\left(a_{H^+} + k_{H^+,Na^+}a_{Na^+}\right)$$

Metallic Function of the Glass Electrode

Before the 1950s, there was no explanation for some important aspects of the behavior of glass electrodes (GE) and the factual reversibility of this behavior. Some authors have refuted the existence of a particular function at GE in such solutions where they do not behave fully as hydrogen electrode, d enying the phenomena of these functions, which were attributed by researchers as an incorrect interpretation of the structural changes in the surface layers of glass; it was mistakenly attributing the changes of EMF element to change in the capacity of GE, and therefore received too large values of pH.

George Eisenman wrote in his retrospective review:

The pioneering studies of Lengiel and Blum were extended by others, who were primarily interested in the existence of Na^+ sensitivity *per se* (i.e., the Na^+ selectivity relative to H^+ only) and in establishing whether or not the electrodes were reversible in the thermodynamic sense. A review of this work is given by Shultz, whose studies and those of Nicolskii and Tolmacheva are noteworthy. In fact, Shultz was the first to demonstrate, by direct comparison with a sodium amalgam electrode, that glasses behave as reversible electrodes for Na^+ at neutral and alkaline pH.

In 1951 Mikhail Schultz, first proved rigorously the thermodynamical reversibility of the Na-function of different glasses in different pH ranges (later the functions for other metal ions) that confirmed the validity of one of the key hypotheses of ion-exchange theory, now the Nikolsky-Shultz-Eisenman thermodynamic ion-exchange theory of GE.

This fact is important because ion-exchange theory was confirmed after a thermodynamically rigorous experimental confirmation of metallic function only. Before, it could be called only as a hypothesis (an epistemological). This opened the way for industrial technology GE, forming ionometry with them, later with membrane electrodes. In the context of "generalized" theory of glass electrodes, Shultz has created a framework for interpreting a mechanism of the influence of diffusion processes in glasses and resin in their electrode properties, giving new quantitative relationship, which take into account the dynamic and energetic characteristics of ion exchangers. Schulz introduced a thermodynamic consideration of processes in the membranes. Considering the different abilities of the dissociation of ionogenic groups of the glasses, his theory allows a rigorous analytical way to connect the electrode properties of glasses and ion-exchange resins with their chemical characteristics.

Range of a pH Glass Electrode

The pH range at constant concentration can be divided into 3 parts:

- Complete realization of general electrode function, where potential depends linearly on pH, realizing an ion-selective electrode for hydronium.

$$E = E^0 + \frac{2.303RT}{F}\mathrm{pH}$$

where F is Faraday's constant.

- Alkali error range - at low concentration of hydrogen ions (high values of pH) contributions of interfering alkali metals (like Li, Na, K) are comparable with the one of hydrogen ions. In this situation dependence of the potential on pH become non-linear.

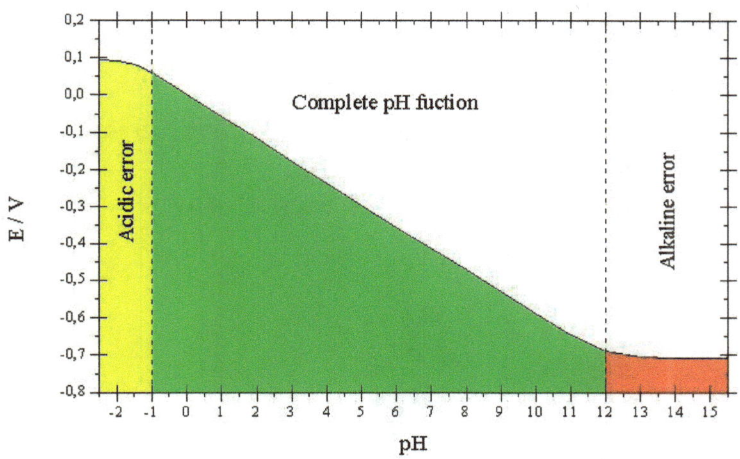

Scheme of the typical dependence E (Volt) – pH for ion-selective electrode.

The effect is usually noticeable at pH > 12, and concentrations of lithium or sodium ions of 0.1 moles per litre or more. Potassium ions usually cause less error than sodium ions.

- Acidic error range – at very high concentration of hydrogen ions (low values of pH) the dependence of the electrode on pH becomes non-linear and the influence of the anions in the solution also becomes noticeable. These effects usually become noticeable at pH < -1.

There are different types of pH glass electrode, some of them have improved characteristics for working in alkaline or acidic media. But almost all electrodes have sufficient properties for working in the most popular pH range from pH = 2 to pH = 12. Special electrodes should be used only for working in aggressive conditions.

Most of text written above is also correct for any ion-exchange electrodes.

Construction

A typical modern pH probe is a combination electrode, which combines both the glass and reference electrodes into one body. The combination electrode consists of the following parts:

- a sensing part of electrode, a bulb made from a specific glass
- internal electrode, usually silver chloride electrode or calomel electrode
- internal solution, usually a pH=7 buffered solution of 0.1 mol/L KCl for pH electrodes or 0.1 mol/L MeCl for pMe electrodes
- when using the silver chloride electrode, a small amount of AgCl can precipitate inside the glass electrode
- reference electrode, usually the same type as 2
- reference internal solution, usually 0.1 mol/L KCl
- junction with studied solution, usually made from ceramics or capillary with asbestos or quartz fiber.
- body of electrode, made from non-conductive glass or plastics.

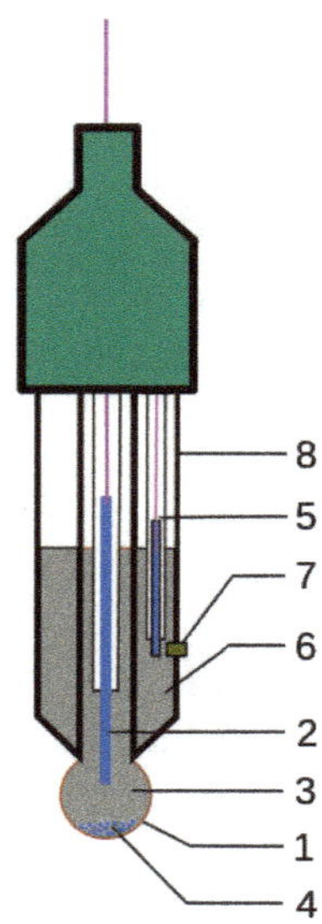

Scheme of typical pH glass electrode.

The bottom of a pH electrode balloons out into a round thin glass bulb. The pH electrode is best thought of as a tube within a tube. The inner tube contains an unchanging 1×10^{-7} mol/L HCl solution. Also inside the inner tube is the cathode terminus of the reference probe. The anodic terminus wraps itself around the outside of the inner tube and ends with the same sort of reference probe as was on the inside of the inner tube. It is filled with a reference solution of KCl and has contact with the solution on the outside of the pH probe by way of a porous plug that serves as a salt bridge.

Galvanic Cell Schematic Representation

This section describes the functioning of two distinct types of electrodes as one unit which combines both the glass electrode and the reference electrode into one body. It deserves some explanation.

This device is essentially a galvanic cell that can be schematically represented as:

Glass electrode || Reference Solution || Test Solution || Glass electrode

Ag(*s*) | AgCl(*s*) | KCl(*aq*) || 1×10^{-7}M H^+ solution || glass membrane || Test Solution || junction || KCl(*aq*) | AgCl(*s*) | Ag(*s*)

In this schematic representation of the galvanic cell, one will note the symmetry between the left and the right members as seen from the center of the row occupied by the "Test Solution" (the solution whose pH must be measured). In other words, the glass membrane and the ceramic junction occupies both the same relative place in each respective electrode (indicative (sensing) electrode or reference electrode). The double "pipe symbol" (||) indicates a diffusive barrier that prevents (glass membrane), or slowing down (ceramic junction), the mixing of the different solutions. By using the same electrodes on the left and right, any potentials generated at the interfaces cancel each other (in principle), resulting in the system voltage being dependent only on the interaction of the glass membrane and the test solution.

The measuring part of the electrode, the glass bulb on the bottom, is coated both inside and out with a ~10 nm layer of a hydrated gel. These two layers are separated by a layer of dry glass. The silica glass structure (that is, the conformation of its atomic structure) is shaped in such a way that it allows Na^+ ions some mobility. The metal cations (Na^+) in the hydrated gel diffuse out of the glass and into solution while H^+ from solution can diffuse into the hydrated gel. It is the hydrated gel, which makes the pH electrode an ion-selective electrode.

H^+ does not cross through the glass membrane of the pH electrode, it is the Na^+ which crosses and allows for a change in free energy. When an ion diffuses from a region of activity to another region of activity, there is a free energy change and this is what the pH meter actually measures. The hydrated gel membrane is connected by Na^+ transport and thus the concentration of H^+ on the outside of the membrane is 'relayed' to the inside of the membrane by Na^+.

All glass pH electrodes have extremely high electric resistance from 50 to 500 MΩ. Therefore, the glass electrode can be used only with a high input-impedance measuring device like a pH meter, or, more generically, a high input-impedance voltmeter which is called an electrometer.

Storage

Between measurements any glass and membrane electrodes should be kept in a solution of its own ion It is necessary to prevent the glass membrane from drying out because the performance is dependent on the existence of a hydrated layer, which forms slowly.

Palladium-Hydrogen Electrode

The palladium-hydrogen electrode (abbreviation: Pd/H_2) is one of the common reference electrodes used in electrochemical study. Most of its characteristics are similar to the standard hydrogen electrode (with platinum). But palladium has one significant feature—the capability to absorb (dissolve into itself) molecular hydrogen.

Electrode operation

Two phases can coexist in palladium when hydrogen is absorbed:

- alpha-phase at hydrogen concentration less than 0.025 atoms per atom of palladium
- beta-phase at hydrogen concentration corresponding to the non-stoichiometric formula $PdH_{0.6}$

The electrochemical behaviour of a palladium electrode in equilibrium with H_3O^+ ions in solution parallels the behaviour of palladium with molecular hydrogen

$$\frac{1}{2}H_2 = H_{ads} = H_{abs}$$

Thus the equilibrium is controlled in one case by the partial pressure or fugacity of molecular hydrogen and in other case—by activity of H^+-ions in solution.

$$E = E^0 + \frac{RT}{F}\ln\frac{a_{H^+}}{(\frac{p_{H2}}{p^0})^{1/2}}$$

When palladium is electrochemically charged by hydrogen, the existence of two phases is manifested by a constant potential of approximately +50 mV compared to the reversible hydrogen electrode. This potential is independent of the amount of hydrogen absorbed over a wide range. This property has been utilized in the construction of a palladium/hydrogen reference electrode. The main feature of such electrode is an absence of non-stop bubbling of molecular hydrogen through the solution as it is absolutely necessary for the standard hydrogen electrode.

Dynamic Hydrogen Electrode

A dynamic hydrogen electrode (DHE) is a reference electrode, more specific a subtype of the standard hydrogen electrodes for electrochemical processes by simulating a reversible hydrogen electrode with an approximately 20 to 40 mV more negative potential.

Principle

A separator in a glass tube connects two electrolytes and a small current is enforced between the cathode and anode.

Applications

- In-situ reference electrode for direct methanol fuel cells
- Proton exchange membrane fuel cells

Nonaqueous Reference Electrodes

While it is convenient to compare between solvents to qualitatively compare systems it is not quantitatively meaningful. Much as pK_a are related between solvents, but not the same, so is the case with E°. While the SHE might seem to be a reasonable reference for nonaqueous work as it turns out the platinum is rapidly poisoned by many solvents including acetonitrile causing uncontrolled drifts in potential. Both the SCE and saturated Ag/AgCl are aqueous electrodes based around saturated aqueous solution. While for short periods it may be possible to use such aqueous electrodes as references with nonaqueous solutions the long-term results are not trustworthy. Using aqueous electrodes introduces undefined, variable, and unmeasurable junction potentials to the cell in the form of a liquid-liquid junction as well as different ionic composition between the reference compartment and the rest of the cell. The best argument against using aqueous reference electrodes with nonaqueous systems, as mentioned earlier, is that potentials measured in different solvents are not directly comparable. For instance, the potential for the $Fc^{0/+}$ couple is sensitive to solvent.

Solvent	**$E_{1/2}$ ($FeCp_2^{0/+}$ vs SCE, 0.1 M NBu_4PF_6 at 298 K)**
CH_3CN	0.40, 0.382
CH_2Cl_2	0.46, 0.475
THF	0.56, 0.547
DMF	0.45, 0.470
acetone	0.48
DMSO	0.435
DME	0.51, 0.580

A Quasi-Reference Electrode (QRE) avoids the issues mentioned above. A QRE with ferrocene or another internal standard, such as cobaltocene or decamethylferrocene, referenced back to ferrocene is ideal for nonaqueous work. Since the early 1960s ferrocene has been gaining acceptance as the standard reference for nonaqueous work for a number of reasons, and in 1984, IUPAC recommended ferrocene (II/III) as a standard redox couple. The preparation of the QRE electrode is simple, allowing for a fresh reference to be prepared with each set of experiments. Since QREs are made fresh, there is also no concern with improper storage or maintenance of the electrode. QREs are also more affordable than other reference electrodes.

To make a quasi-reference electrode (QRE):

- Insert a piece of silver wire into concentrated HCl then allow the wire to dry on a lint-free cleaning cloth. This forms an insoluble layer of AgCl on the surface of the electrode and gives you an Ag/AgCl wire. Repeat dipping every few months or if the QRE starts to drift.
- Obtain a Vycor glass frit (4 mm diameter) and glass tubing of similar diameter. Attach Vycor glass frit to the glass tubing with heat shrink Teflon tubing.
- Rinse then fill the clean glass tube with supporting electrolyte solution and insert Ag/AgCl wire.

- The ferrocene (II/III) couple should lie around 400 mV versus this Ag/AgCl QRE in an acetonitrile solution. This potential will vary up to 200 mV with specific undefined conditions, thus adding an internal standard such as ferrocene at some point during the experiment is always necessary.

Pseudo Reference Electrodes

A pseudo reference electrode is a term that is not well defined and borders on having multiple meanings since *pseudo* and *quasi* are often used interchangeably. They are a class of electrodes named pseudo-reference electrodes because they do not maintain a constant potential but vary predictably with conditions. If the conditions are known, the potential can be calculated and the electrode can be used as a reference. Most electrodes work over a limited range of conditions, such as pH or temperature, outside of this range the electrodes behavior becomes unpredictable. The advantage of a pseudo-reference electrode is that the resulting variation is factored into the system allowing researchers to accurately study systems over a wide range of conditions.

Yttria-stabilized zirconia (YSZ) membrane electrodes were developed with a variety of redox couples, e.g., Ni/NiO. Their potential depends on pH. When the pH value is known, these electrodes can be employed as a reference with notable applications at elevated temperatures.

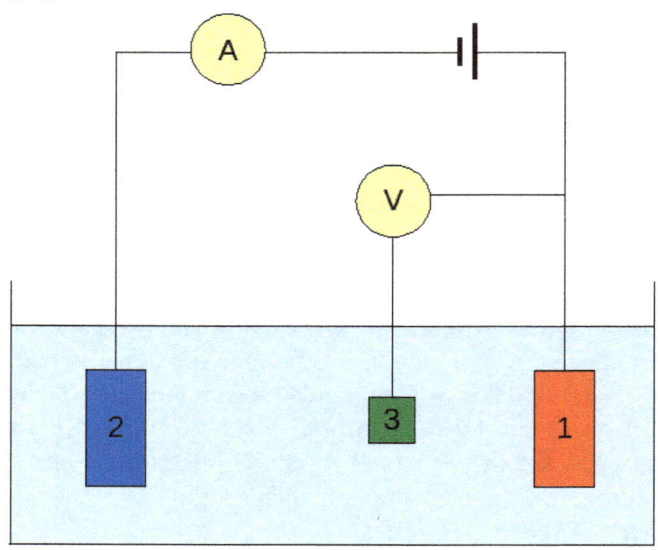

Three-electrode setup for measurement of electrode potential.

In case of non-zero net current on the electrode, it is essential to minimize the ohmic IR-drop in the electrolyte, e.g., by positioning the reference electrode near the surface of the working electrode (e.g., see Luggin capillary), or by using a supporting electrolyte of sufficiently high conductivity. The potential measurements are performed with the positive terminal of the electrometer connected to working electrode and the negative terminal to the reference electrode.

SIGN CONVENTIONS

Historically, two conventions for sign for the electrode potential have formed:

- Convention "Nernst-Lewis-Latimer" (sometimes referred to as "American");
- Convention "Gibbs-Ostwald-Stockholm" (sometimes referred to as "European").

In 1953 in Stockholm IUPAC recognized that either of the conventions is permissible; however, it unanimously recommended that only the magnitude expressed according to the convention (2) be called "the electrode potential". To avoid possible ambiguities, the electrode potential thus defined can also be referred to as *Gibbs-Stockholm electrode potential*.

In both conventions, the EMF of the half-cell is referred to the standard hydrogen electrode as the zero point.

According to the Nernst-Lewis-Latimer convention (1), the sign of the potential is determined by the change of the sign of the standard Gibbs free energy of the half-cell reaction. This reaction transfers *nF* of the positive electric charge across the phase boundary from the left to the right of the half-cell, as shown in the half-cell diagram. The relation is written as:

$$\Delta G = -nFE$$

where: ΔG - the standard Gibbs energy for the half-cell reaction, n - number of electrons transferred across the across the phase boundary in the above reaction, F - the Faraday constant, E - the half-cell potential according to the Nernst-Lewis-Latimer convention.

Therefore, according to the convention (1), the value of EMF is bivariant (with two variants), i.e., reversing the electrode reaction, and reversing the corresponding cell diagram, will cause the change of sign of ΔG and E, even though the polarity of the cell measured with a direct-current voltmeter does not necessarily reverse. This convention has been historically (sometimes still is) used to express anodic oxidation potentials or cathodic reduction potentials of half-cells. The reduction potentials expressed according to (1) have the same sign as those expressed according to the convention (2).

According to the Gibbs-Stockholm convention (2), the sign of the electrode potential is determined by the polarity of the cell terminal measured against the standard hydrogen electrode using an instrument for DC-potential measurements, when this circuit is closed and balanced (i.e., the current on the measurement circuit is nil). In convention (2), the value of the measured EMF is invariant of the direction of the infinitesimally small changes of the electric current that may be allowed on the measured electrode and render it either cathodic or anodic. Therefore, the value of E does not change with the direction of the (very small) current, even though ΔG of the corresponding reaction is bivariant (can be of either positive or negative sign).

Potential Difference of a Cell Assembled of Two Electrodes

Potential of a cell assembled of two electrodes can be determined from the two individual electrode potentials using:

$$\Delta V_{cell} = E_{red,cathode} - E_{red,anode}$$

or, equivalently,

$$\Delta V_{cell} = E_{red,cathode} + E_{oxy,anode}$$

This follows from the IUPAC definition of the electric potential difference of a galvanic cell, according to which the electric potential difference of a cell is the difference of the potentials of the electrodes on the right and the left of the galvanic cell. When ΔV_{cell} is positive, then positive electrical charge flows through the cell from the left electrode (anode) to the right electrode (cathode).

STANDARD ELECTRODE POTENTIAL

In electrochemistry, the standard electrode potential, abbreviated $E°$ or E (with a superscript plimsoll character, pronounced "standard" or "nought"), is the measure of individual potential of a reversible electrode at standard state, which is with solutes at an effective concentration of 1 mol dm^{-3}, and gases at a pressure of 1 atm. The reduction potential is an intensive property. The values are most often tabulated at 25 °C. The basis for an electrochemical cell such as the galvanic cell is always a redox reaction which can be broken down into two half-reactions: oxidation at anode (loss of electron) and reduction at cathode (gain of electron). Electricity is generated due to electric potential difference between two electrodes. This potential difference is created as a result of the difference between individual potentials of the two metal electrodes with respect to the electrolyte. (Reversible electrode is an electrode that owes its potential to changes of a reversible nature, in contrast to electrodes used in electroplating which are destroyed during their use.)

Although the overall potential of a cell can be measured, there is no simple way to accurately measure the electrode/electrolyte potentials in isolation. The electric potential also varies with temperature, concentration and pressure. Since the oxidation potential of a half-reaction is the negative of the reduction potential in a redox reaction, it is sufficient to calculate either one of the potentials. Therefore, standard electrode potential is commonly written as standard reduction potential.

Calculation of Standard Electrode Potentials

The electrode potential cannot be obtained empirically. The galvanic cell potential results from a *pair* of electrodes. Thus, only one empirical value is available in a pair of electrodes and it is not possible to determine the value for each electrode in the pair using the empirically obtained galvanic cell potential. A reference electrode, standard hydrogen electrode (SHE), for which the potential is *defined* or agreed upon by convention, needed to be established. In this case SHE is set to 0.00 V and any electrode, for which the electrode potential is not yet known, can be paired with SHE—to form a galvanic cell—and the galvanic cell potential gives the unknown electrode's potential. Using this process, any electrode with an unknown potential can be paired with either the SHE or another electrode for which the potential has already been derived and that unknown value can be established.

Since the electrode potentials are conventionally defined as reduction potentials, the sign of the potential for the metal electrode being oxidized must be reversed when calculating the overall cell potential. Note that the electrode potentials are independent of the number of electrons transferred —they are expressed in volts, which measure energy per electron transferred—and so the two electrode potentials can be simply combined to give the overall *cell* potential even if different numbers of electrons are involved in the two electrode reactions.

For practical measurements, the electrode in question is connected to the positive terminal of the electrometer, while SHE is connected to the negative terminal.

Standard Reduction Potential Table

The larger the value of the standard reduction potentials, the easier it is for the element to be reduced (accept electrons); in other words, they are better oxidizing agents. For example, F_2has 2.87 V and Li^+ has −3.05 V. F reduces easily and is therefore a good oxidizing agent. In contrast, $Li_{(s)}$ would rather undergo oxidation (hence a good reducing agent). Thus Zn^{2+}whose standard reduction potential is −0.76 V can be oxidized by any other electrode whose standard reduction potential is greater than −0.76 V (e.g. H^+(0 V), Cu^{2+}(0.16 V), F_2(2.87 V)) and can be reduced by any electrode with standard reduction potential less than −0.76 V (e.g. H_2(−2.23 V), Na^+(−2.71 V), Li^+(−3.05 V)).

In a galvanic cell, where a spontaneous redox reaction drives the cell to produce an electric potential, Gibbs free energy $\Delta G°$ must be negative, in accordance with the following equation:

$$\Delta G°_{cell} = -nFE°_{cell}$$

where n is number of moles of electrons per mole of products and F is the Faraday constant, ~96485 C/mol. As such, the following rules apply:

If $E°_{cell} > 0$, then the process is spontaneous (galvanic cell)

If $E°_{cell} < 0$, then the process is nonspontaneous (electrolytic cell)

Thus in order to have a spontaneous reaction ($\Delta G° < 0$), $E°_{cell}$ must be positive, where:

$$E°_{cell} = E°_{cathode} - E°_{anode}$$

where $E°_{anode}$ is the standard potential at the anode and $E°_{cathode}$ is the standard potential at the cathode as given in the table of standard electrode potential.

ABSOLUTE ELECTRODE POTENTIAL

Absolute electrode potential, in electrochemistry, according to an IUPAC definition, is the electrode potential of a metal measured with respect to a universal reference system (without any additional metal–solution interface).

Definition

According to a more specific definition presented by Trasatti, the absolute electrode potential is the difference in electronic energy between a point inside the metal (Fermi level) of an electrode and a point outside the electrolyte in which the electrode is submerged (an electron at rest in vacuum).

This potential is difficult to determine accurately. For this reason, standard hydrogen electrode is typically used for reference potential. The absolute potential of the SHE is 4.44 ± 0.02 V at 25 °C. Therefore, for any electrode at 25 °C:

$$E^{M}_{(abs)} = E^{M}_{(SHE)} + (4.44 \pm 0.02)\ V$$

where:

E is electrode potential

V is volt

M denotes the electrode made of metal M

(abs) denotes the absolute potential

(SHE) denotes the electrode potential relative to the standard hydrogen electrode.

A different definition for the absolute electrode potential (also known as absolute half-cell potential and single electrode potential) has also been discussed in the literature. In this approach, one first defines an isothermal absolute single-electrode process (or absolute half-cell process.) For example, in the case of a generic metal being oxidized to form a solution-phase ion, the process would be

$$M_{(metal)} \rightarrow M^{+}_{(solution)} + e^{-}_{(gas)}$$

For the hydrogen electrode, the absolute half-cell process would be

$$1/2H_{2\ (gas)} \rightarrow H^{+}_{(solution)} + e^{-}_{(gas)}$$

Other types of absolute electrode reactions would be defined analogously.

In this approach, all three species taking part in the reaction, including the electron, must be placed in thermodynamically well-defined states. All species, including the electron, are at the same temperature, and appropriate standard states for all species, including the electron, must be fully defined. The absolute electrode potential is then defined as the Gibbs free energy for the absolute electrode process. To express this in volts one divides the Gibb's free energy by the negative of Faraday's constant.

Rockwood's approach to absolute-electrode thermodynamics is easily expendable to other thermodynamic functions. For example, the absolute half-cell entropy has been defined as the entropy of the absolute half-cell process defined above. An alternative definition of the absolute half-cell entropy has recently been published by Fang et al. who define it as the entropy of the following reaction (using the hydrogen electrode as an example):

$$1/2H_{2\ (gas)} \rightarrow H^{+}_{(solution)} + e^{-}_{(metal)}$$

This approach differs from the approach described by Rockwood in the treatment of the electron, i.e. whether it is placed in the gas phase or in the metal.

Determination

The basis for determination of the absolute electrode potential under the Trasatti definition is given by the equation:

$$E^M(\mathbf{abs}) = \phi^M + \Delta_S^M \psi$$

where:

E^M(abs) is the absolute potential of the electrode made of metal M

ϕ^M is the electron work function of metal M

$\Delta_S^M \psi$ is the contact (Volta) potential difference at the metal(M)–solution(S) interface.

For practical purposes, the value of the absolute electrode potential of the standard hydrogen electrode is best determined with the utility of data for an ideally-polarizable mercury (Hg) electrode:

$$E^{\ominus}(\mathbf{H^+/H_2})(\mathbf{abs}) = \phi^{\mathbf{Hg}} + \Delta_S^{\mathbf{Hg}} \psi_{\sigma=0}^{\ominus} - E_{\sigma=0}^{\mathbf{Hg}}(\mathbf{SHE})$$

where:

$E^{\ominus}(\mathbf{H^+/H_2})(\mathbf{abs})$ is the absolute standard potential of the hydrogen electrode

$\sigma = 0$ denotes the condition of the point of zero charge at the interface.

The types of physical measurements required under the Rockwood definition are similar to those required under the Trasatti definition, but they are used in a different way, e.g. in Rockwood›s approach they are used to calculate the equilibrium vapor pressure of the electron gas. The numerical value for the absolute potential of the standard hydrogen electrode one would calculate under the Rockwood definition is sometimes fortuitously close to the value one would obtain under the Trasatti definition. This near-agreement in the numerical value depends on the choice of ambient temperature and standard states, and is the result of the near-cancellation of certain terms in the expressions. For example, if a standard state of one atmosphere ideal gas is chosen for the electron gas then the cancellation of terms occurs at a temperature of 296 K, and the two definitions give an equal numerical result. At 298.15 K a near-cancellation of terms would apply and the two approaches would produce nearly the same numerical values. However, there is no fundamental significance to this near agreement because it depends on arbitrary choices, such as temperature and definitions of standard states.

REFERENCES

1. IUPAC. Compendium of Chemical Terminology, 2nd ed. (the "Gold Book"). Compiled by A. D. McNaught and A. Wilkinson. Blackwell Scientific Publications, Oxford (1997). XML on-line corrected version: http://goldbook.iupac.org (2006-) created by M. Nic, J. Jirat, B. Kosata; updates compiled by A. Jenkins. ISBN 0-9678550-9-8. doi:10.1351/goldbook. Entry: "Electrode Potential"

2. C.A. Hamel, "The Encyclopedia of Electrochemistry", Reinhold Publishing Corporation, New York-Chapman & Hall Ltd., London, 1964, s. 429-431.
3. P. van Rysselberghe, "Bericht der Kommission für elektrochemische Nomenklatur und Definitionen", Z. Electrochem., 58 (1954), 530-535.
4. IUPAC Gold Book. Definition of the potential difference of a galvanic cell. http://goldbook.iupac.org/E01934.html
5. Zumdahl, Steven S., Zumdahl, Susan A (2000) Chemistry (5th ed.), Houghton Mifflin Company. ISBN 0-395-98583-8
6. Atkins, Peter, Jones, Loretta (2005) Chemical Principles (3rd ed.), W.H. Freeman and Company. ISBN 0-7167-5701-X
7. Zu, Y, Couture, MM, Kolling, DR, Crofts, AR, Eltis, LD, Fee, JA, Hirst, J (2003) Biochemistry, 42, 12400-12408
8. Shuttleworth, SJ (1820) Electrochemistry (50th ed.), Harper Collins.
9. IUPAC Gold Book - absolute electrode potential
10. Sergio Trasatti, "The Absolute Electrode Potential: an Explanatory Note (Recommendations 1986)", International Union of Pure and Applied Chemistry, Pure & AppL Chem., Vol. 58, No. 7, pp. 955–66, 1986. http://www.iupac.org/publications/pac/1986/pdf/5807x0955.pdf (pdf)
11. Alan L. Rockwood, "Absolute half-cell thermodynamics: Electrode potential", Physical Review A, Vol 33, No. 1, pp. 554–59, 1986.
12. Alan L. Rockwood, "Absolute half-cell entropy", Physical Review A, vol. 36, No. 3, pp. 1525–26, 1987.
13. Zheng Fang, Shaofen Wang, Zhenghua Zhang, and Guanzhou Qiu, "The electrochemical Peltier heat of the standard hydrogen electrode reaction", Thermochimica Acta, Vol. 473, pp. 40–44, 2008.

CHAPTER 3

Electrode Potential

Chapter Contents

Electrokinetic phenomena is a group of effects that are observed in disperse systems and in capillaries and that entail either the movement of one phase relative to another under the influence of an applied electric field or the production of a potential difference in the direction of the relative movement of phases that is due to mechanical forces. Electroosmosis and electrophoresis are the electrokinetic phenomena of the first type; the production of a sedimentation potential, or the Dorn effect, and the production of a streaming potential are the electrokinetic phenomena of the second type.

Electrokinetic phenomena result from the existence, at a phase boundary, of excess charges that are grouped as two oppositely charged layers called an electrical double layer. An electric field applied in the direction parallel to a phase boundary causes the displacement of an ionic layer relative to another, resulting in the relative displacement of the phases, that is, in electroosmosis or electrophoresis. Effects that are the converse of electroosmosis or electrophoresis are similarly observed in the flow of a liquid or in the sedimentation of particles from a disperse phase. Such effects entail the relative movement of ionic layers or the spatial separation of charges in the direction in which the phases move. In other words, they entail the production of a streaming potential or the production of a sedimentation potential, respectively.

Any electrokinetic phenomenon may be used to determine the zeta potential, or electrokinetic potential, which is denoted ζ. In such a determination, the surface conductivity caused by the freely movable charges of an electrical double layer is assumed to be greater than the volume conductivity of a system.

INTRODUCTION

Electrokinetic phenomena are a family of several different effects that occur in heterogeneous fluids, or in porous bodies filled with fluid, or in a fast flow over a flat surface. The term heterogeneous here means a fluid containing particles. Particles can be solid, liquid or gas bubbles with sizes on the scale of a micrometer or nanometer. There is a common source of all these effects — the so-called interfacial 'double layer' of charges. Influence of an external force on the diffuse layer generates tangential motion of a fluid with respect to an adjacent charged surface. This force might be electric, pressure gradient, concentration gradient, or gravity. In addition, the moving phase might be either continuous fluid or dispersed phase.

FAMILY OF ELECTROKINETIC PHENOMENA

Various combinations of the driving force and moving phase determine various electrokinetic effects. According to J.Lyklema, the complete family of electrokinetic phenomena includes:

- electrophoresis, as motion of particles under influence of electric field;
- electro-osmosis, as motion of liquid in porous body under influence of electric field;
- diffusiophoresis, as motion of particles under influence of a chemical potential gradient;
- capillary osmosis, as motion of liquid in porous body under influence of the chemical potential gradient;
- sedimentation potential, as electric field generated by sedimenting colloid particles;
- streaming potential/current, as either electric potential or current generated by fluid moving through porous body, or relative to flat surface;
- colloid vibration current, as electric current generated by particles moving in fluid under influence of ultrasound;
- electric sonic amplitude, as ultrasound generated by colloidal particles in oscillating electric field.

Electrophoresis

Electrophoresis is the motion of dispersed particles relative to a fluid under the influence of a spatially uniform electric field. This electrokinetic phenomenon was observed for the first time in 1807 by Ferdinand Frederic Reuss (Moscow State University), who noticed that the application of a constant electric field caused clay particles dispersed in water to migrate. It is ultimately caused by the presence of a charged interface between the particle surface and the surrounding fluid. It is the basis for a number of analytical techniques used in chemistry for separating molecules by size, charge, or binding affinity.

Electrophoresis of positively charged particles (cations) is called cataphoresis, while electrophoresis of negatively charged particles (anions) is called anaphoresis. Electrophoresis is a technique used in laboratories in order to separate macromolecules based on size. The technique applies a negative charge so proteins move towards a positive charge. This is used for both DNA and RNA analysis. Polyacrylamide gel electrophoresis (PAGE) has a clearer resolution than agarose and is more suitable for quantitative analysis. In this technique DNA foot-printing can identify how proteins bind to DNA. It can be used to separate proteins by size, density and purity. It can also be used for plasmid analysis, which develops our understanding of bacteria becoming resistant to antibiotics.

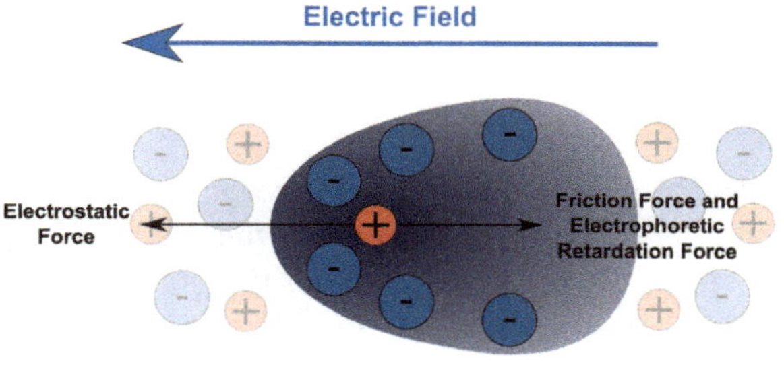

Illustration of electrophoresis.

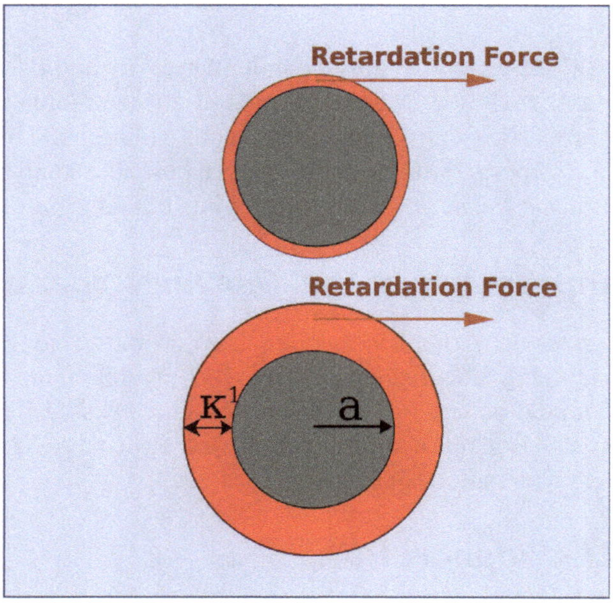

Illustration of electrophoresis retardation.

History of Electrophoresis

The history of electrophoresis begins in earnest with the work of Arne Tiselius in the 1931, and new separation processes and chemical analysis techniques based on electrophoresis continue to be developed into the 21st century. Tiselius, with support from the Rockefeller Foundation, developed the "Tiselius apparatus" for moving boundary electrophoresis, which was described in 1937 in the well-known paper "A New Apparatus for Electrophoretic Analysis of Colloidal Mixtures". The method spread slowly until the advent of effective zone electrophoresis methods in the 1940s and 1950s, which used filter paper or gels as supporting media. By the 1960s, increasingly sophisticated gel electrophoresis methods made it possible to separate biological molecules based on minute physical and chemical differences, helping to drive the rise of molecular biology. Gel electrophoresis and related techniques became the basis for a wide range of biochemical methods, such as protein fingerprinting, Southern blot and similar blotting procedures, DNA sequencing, and many more.

Electrophoresis before Tiselius

Early work with the basic principle of electrophoresis dates to the early 19th century, based on Faraday's laws of electrolysis proposed in the late 18th century and other early electrochemistry. Experiments by Johann Wilhelm Hittorf, Walther Nernst, and Friedrich Kohlrausch to measure the properties and behavior of small ions moving through aqueous solutions under the influence of an electric field led to general mathematical descriptions of the electrochemistry of aqueous solutions. Kohlrausch created equations for varying concentrations of charged particles moving through solution, including sharp moving boundaries of migrating particles. By the beginning of the 20th century, electrochemists had found that such moving boundaries of charged particles could be created with U-shaped glass tubes.

Methods of optical detection of moving boundaries in liquids had been developed by August Toepler in the 1860s; Toepler measured the schlieren (*shadows*) or slight variations in optical properties that in inhomogeneous solutions. This method combined with the theoretical and experimental methods for creating and analysing charged moving boundaries would form the basis of Tiselius's moving boundary electrophoresis method.

Development and Spread of the Tiselius Apparatus

The apparatus designed by Arne Tiselius enabled a range of new applications of electrophoresis in analyzing chemical mixtures. Its development, significantly funded by the Rockefeller Foundation, was an extension of Tiselius's earlier PhD studies. With more assistance from the Rockefeller Foundation, the expensive Tiselius apparatus was built at a number of major centers of chemical research.

Rise of Zone Electrophoresis

By the late 1940s, new electrophoresis methods were beginning to address some of the shortcomings of the moving boundary electrophoresis of the Tiselius apparatus, which was not capable of completely separating electrophoretically similar compounds. Rather than charged molecules moving freely through solutions, the new methods used solid or gel matrices to separate compounds into discrete and stable bands (zones); in 1950 Tiselius dubbed these methods "zone electrophoresis".

Zone electrophoresis found widespread application in biochemistry after Oliver Smithies introduced starch gel as an electrophoretic substrate in 1955. Starch gel (and later polyacrylamide and other gels) enabled the efficient separation of proteins, making it possible with relatively simple technology to analyze complex protein mixtures and identify minute differences in related proteins.

Widespread Application

Since the 1950s, electrophoresis methods have diversified considerably, and new methods and applications are still being developed as Affinity electrophoresis, Capillary electrophoresis, Electroblotting, Electrophoretic mobility shift assay, Isotachophoresis, Pulsed-field gel electrophoresis, and Preparative electrophoresis.

Theory

Suspended particles have an electric surface charge, strongly affected by surface adsorbed species, on which an external electric field exerts an electrostatic Coulomb force. According to the double layer theory, all surface charges in fluids are screened by a diffuse layer of ions, which has the same absolute charge but opposite sign with respect to that of the surface charge. The electric field also exerts a force on the ions in the diffuse layer which has direction opposite to that acting on the surface charge. This latter force is not actually applied to the particle, but to the ions in the diffuse layer located at some distance from the particle surface, and part of it is transferred all the way to the particle surface through viscous stress. This part of the force is also called electrophoretic retardation force. When the electric field is applied and the charged particle to be analyzed is at steady movement through the diffuse layer, the total resulting force is zero :

$$F_{tot} = 0 = F_{el} + F_f + F_{ret}$$

Considering the drag on the moving particles due to the viscosity of the dispersant, in the case of low Reynolds number and moderate electric field strength E, the drift velocity of a dispersed particle v is simply proportional to the applied field, which leaves the electrophoretic mobility μ_e defined as:

$$\mu_e = \frac{v}{E}.$$

The most well known and widely used theory of electrophoresis was developed in 1903 by Smoluchowski:

$$\mu_e = \frac{\varepsilon_r \varepsilon_0 \zeta}{\eta},$$

where ε_r is the dielectric constant of the dispersion medium, ε_0 is the permittivity of free space ($C^2\ N^{-1}\ m^{-2}$), η is dynamic viscosity of the dispersion medium (Pa s), and ζ is zeta potential (i.e., the electrokinetic potential of the slipping plane in the double layer).

The Smoluchowski theory is very powerful because it works for dispersed particles of any shape at any concentration. It has limitations on its validity. It follows, for instance, because it does not include Debye length κ^{-1}. However, Debye length must be important for electrophoresis, as follows immediately from the Figure on the right. Increasing thickness of the double layer (DL) leads to removing the point of retardation force further from the particle surface. The thicker the DL, the smaller the retardation force must be.

Detailed theoretical analysis proved that the Smoluchowski theory is valid only for sufficiently thin DL, when particle radius a is much greater than the Debye length:

$$a\kappa \gg 1.$$

This model of "thin double layer" offers tremendous simplifications not only for electrophoresis theory but for many other electrokinetic theories. This model is valid for most aqueous systems, where the Debye length is usually only a few nanometers. It only breaks for nano-colloids in solution with ionic strength close to water.

The Smoluchowski theory also neglects the contributions from surface conductivity. This is expressed in modern theory as condition of small Dukhin number:

$$Du \ll 1$$

In the effort of expanding the range of validity of electrophoretic theories, the opposite asymptotic case was considered, when Debye length is larger than particle radius:

$$a\kappa < 1.$$

Under this condition of a "thick double layer", Hückel predicted the following relation for electrophoretic mobility:

$$\mu_e = \frac{2\varepsilon_r \varepsilon_0 \zeta}{3\eta}.$$

This model can be useful for some nanoparticles and non-polar fluids, where Debye length is much larger than in the usual cases.

There are several analytical theories that incorporate surface conductivity and eliminate the restriction of a small Dukhin number, pioneered by Overbeek. and Booth. Modern, rigorous theories valid for any Zeta potential and often any a stem mostly from Dukhin–Semenikhin theory. In the thin double layer limit, these theories confirm the numerical solution to the problem provided by O'Brien and White.

Electro-Osmosis

Electroosmotic flow (or electro-osmotic flow, often abbreviated EOF; synonymous with electroosmosis or electroendosmosis) is the motion of liquid induced by an applied potential across a porous material, capillary tube, membrane, microchannel, or any other fluid conduit. Because electroosmotic velocities are independent of conduit size, as long as the electrical double layer is much smaller than the characteristic length scale of the channel, electroosmotic flow will have little effect. Electroosmotic flow is most significant when in small channels. Electroosmotic flow is an essential component in chemical separation techniques, notably capillary electrophoresis. Electroosmotic flow can occur in natural unfiltered water, as well as buffered solutions.

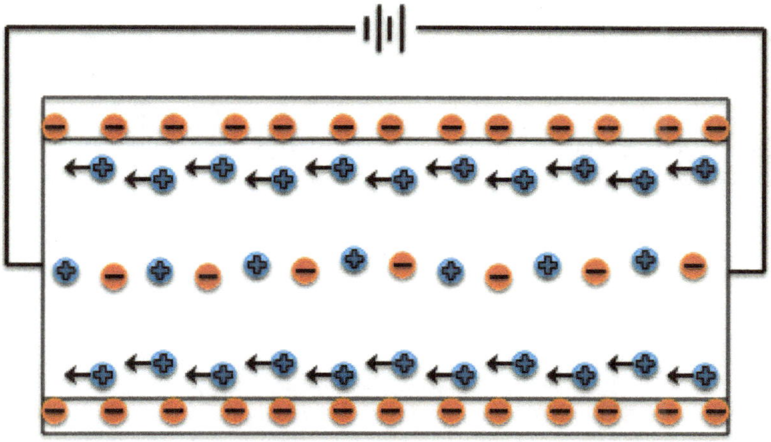

Electroosmotic flow schematic.

History

Electroosmotic flow was first reported in 1809 by F. F. Reuss in the Proceedings of the Imperial Society of Naturalists of Moscow. He showed that water could be made to flow through a plug of clay by applying an electric voltage. Clay is composed of closely packed particles of silica and other minerals, and water flows through the narrow spaces between these particles just as it would through a narrow glass tube. Any combination of an electrolyte (a fluid containing dissolved ions) and an insulating solid would generate electro-osmotic flow, though for water/silica the effect is particularly large. Even so, flow speeds are typically only a few millimeters per second.

Cause

Electroosmotic flow is caused by the Coulomb force induced by an electric field on net mobile electric charge in a solution. Because the chemical equilibrium between a solid surface and an electrolyte solution typically leads to the interface acquiring a net fixed electrical charge, a layer of mobile ions, known as an electrical double layer or Debye layer, forms in the region near the interface. When an electric field is applied to the fluid (usually via electrodes placed at inlets and outlets), the net charge in the electrical double layer is induced to move by the resulting Coulomb force. The resulting flow is termed electroosmotic flow.

Description

The resulting flow from applying a voltage is a plug flow. Unlike a parabolic profile flow generated from a pressure differential, a plug flow's velocity profile is approximately planar, with slight variation near the electric double layer. This offers significantly less deleterious dispersive effects and can be controlled without valves, offering a high performance method for fluid separation, although many complex factors prove this control to be difficult. Because of difficulties measuring and monitoring flow in micro fluidic channels, primarily disrupting the flow pattern, most analysis is done through numerical methods and simulation. Electroosmotic flow through micro channels can be modeled after the Navier-Stokes equation with the driving force deriving from the electric field and not the pressure differential. Thus it is governed by the continuity equation

$$\nabla \cdot \mathbf{U} = 0$$

and momentum

$$\rho \frac{D\mathbf{U}}{Dt} = -\nabla p + \mu \nabla^2 \mathbf{U} + \rho_e \nabla (\psi + \phi),$$

where $\mathbf{U}$ is the velocity vector, ρ is the density of the fluid, D/Dt is the material derivative, μ is the viscosity of the fluid, ρ_e is the electric charge density, Φ is the applied electric field, and ψ is the electric field due to the zeta potential at the walls.

Laplace's equation can describe the external electric field

$$\nabla^2 \phi = 0,$$

while the potential within the electric double layer is governed by

$$\nabla^2 \psi = \frac{-\rho_e}{\epsilon \epsilon_0},$$

where ε is the dielectric constant of the electrolyte solution and ε_0 is the vacuum permittivity. This equation can be further simplified using the Debye-Hückel approximation

$$\nabla^2 \psi = k^2 \psi,$$

where $1 / k$ is the Debye Length, used to describe the characteristic thickness of the electric double layer. The equations for potential field within the double layer can be combined as

$$\rho_e = -\epsilon \epsilon_0 k^2 \psi.$$

Applications

Electroosmotic flow is commonly used in microfluidic devices, soil analysis and processing, and chemical analysis, all of which routinely involve systems with highly charged surfaces, often of oxides. One example is capillary electrophoresis, in which electric fields are used to separate chemicals according to their electrophoretic mobility by applying an electric field to a narrow capillary, usually made of silica. In electrophoretic separations, the electroosmotic flow affects the elution time of the analytes.

It is projected that micro fluidic devices utilizing electroosmotic flow will have great application with medical research. Once controlling this flow is better understood and implemented being able to separate fluids on the atomic level will be a vital component for drug dischargers. Mixing fluids at the micro scale is currently troublesome. It is believed that electrically controlling fluids will be the method in which small fluids are mixed.

A controversially use of electro-osmotic systems is being done to control rising damp in structure of a building. Whilst there is little evidence to suggest that these systems can be useful in moving salts in walls, such systems are claimed to be specially efficient on structures with very thick walls. However some claim that there is no scientific base for those systems, and cite several examples for their failure.

Physics

In fuel cells, electro-osmosis causes protons moving through a proton exchange membrane (PEM) to drag water molecules from one side (anode) to the other (cathode).

Vascular Plant Biology

In vascular plant biology, electro-osmosis is also used as an alternative or supplemental explanation for the movement of polar liquids via the phloem that differs from the cohesion-tension theory supplied in the mass flow hypothesis and others, such as cytoplasmic streaming. Companion cells are involved in the "cyclic" withdrawal of ions (K^+) from sieve tubes, and their secretion parallel to their position of withdrawal between sieve plates, resulting in polarisation of sieve plate elements alongside potential difference in pressure, and results in polar water molecules and other solutes present moved upward through the phloem.

In 2003, St Petersburg University graduates applied direct electric current to 10 mm segments of mesocotyls of maize seedlings alongside one-year linden shoots; electrolyte solutions present in the tissues moved toward the cathode that was in place, suggesting that electro-osmosis might play a role in solution transport through conductive plant tissues.

Disadvantages

Maintaining an electric field in an electrolyte requires Faradaic reactions to occur at the anode and cathode. This is typically electrolysis of water, which generates hydrogen peroxide, hydrogen ions (acid) and hydroxide (base) as well as oxygen and hydrogen gas bubbles. The hydrogen peroxide and/or pH changes generated can adversely affect biological cells and biomolecules such as proteins, while gas bubbles tend to "clog" microfluidic systems. These problems can be alleviated by using alternative electrode materials such as conjugated polymers which can undergo the Faradaic reactions themselves, dramatically reducing electrolysis.

Diffusiophoresis

Diffusiophoresis is a spontaneous motion of dispersed particles in a fluid induced by a diffusion gradient (also called concentration gradient) of molecular substances that are dissolved in the fluid. This gradient affects structure of the particles in an interfacial double layer and causes sliding motion of the fluid relative to the particle surface.

SEDIMENTATION POTENTIAL

Sedimentation Potential occurs when dispersed particles move under the influence of either gravity or centrifugation in a medium. This motion disrupts the equilibrium symmetry of the particle's double layer. While the particle moves, the ions in the electric double layer lag behind due to the liquid flow. This causes a slight displacement between the surface charge and the electric charge of the diffuse layer. As a result, the moving particle creates a dipole moment. The sum of all of the dipoles generates an electric field which is called *sedimentation potential*. It can be measured with an open electrical circuit, which is also called sedimentation current.

Surface Energy

Background Related to Phenomenon

Electrokinetic phenomena are a family of several different effects that occur in heterogeneous fluids or in porous bodies filled with fluid. The sum of these phenomena deals with the effect on a particle from some outside resulting in a net electrokinetic effect.

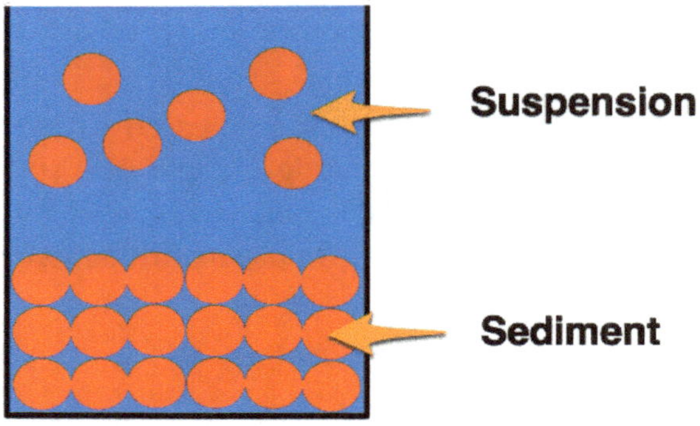

Sedimentation Potential.

The common source of all these effects stems from the interfacial 'double layer' of charges. Particles influenced by an external force generate tangential motion of a fluid with respect to an adjacent charged surface. This force may consist of electric, pressure gradient, concentration gradient, gravity. In addition, the moving phase might be either the continuous fluid or dispersed phase.

Sedimentation potential is the field of electrokinetic phenomena dealing with the generation of an electric field by sedimenting colloid particles.

History of Sedimentation Potential Models

This phenomenon was first discovered by Dorn in 1879. He observed that a vertical electric field had developed in a suspension of glass beads in water, as the beads were settling. This was the origin of sedimentation potential, which is often referred to as the Dorn effect.

Smoluchowski built the first models to calculate the potential in the early 1900s. Booth created a general theory on sedimentation potential in 1954 based on Overbeek's 1943 theory on electrophoresis. In 1980, Stigter extended Booth's model to allow for higher surface potentials. Ohshima created a model based on O'Brien and White 's 1978 model used to analyze the sedimentation velocity of a single charged sphere and the sedimentation potential of a dilute suspension.

Generation of a Potential

As a charged particle moves through a gravitational force or centrifugation, an electric potential is induced. While the particle moves, ions in the electric double layer lag behind creating a net dipole moment behind due to liquid flow. The sum of all dipoles on the particle is what causes sedimentation potential. Sedimentation potential has the opposite effect compared to electrophoresis where an electric field is applied to the system. Ionic conductivity is often referred to when dealing with sedimentation potential.

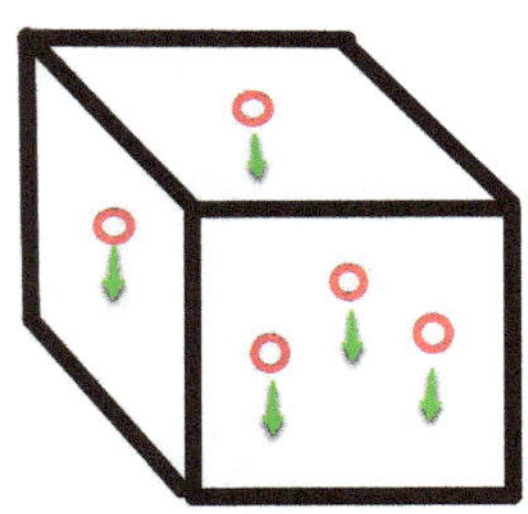

Macroscopic Diagram of Sedementation.

The following relation provides a measure of the sedimentation potential due to the settling of charged spheres. First discovered by Smoluchowski in 1903 and 1921. This relationship only holds true for non-overlapping electric double layers and for dilute suspensions. In 1954, Booth proved that this idea held true for Pyrex glass powder settling in a KCl solution. From this relation, the sedimentation potential, E_S, is independent of the particle radius and that $E_S \rightarrow 0$, $\Phi_p \rightarrow 0$ (a single particle).

$$E_s = -\frac{\varepsilon\zeta(\rho - \rho_0)\phi_p g}{\sigma^{\infty}\eta}$$

Smoluchowski's sedimentation potential is defined where ε_0 is the permitivity of free space, D the dimensionless dielectric constant, ξ the zeta potential, g the acceleration due to gravity, Φ the particle volume fraction, ρ the particle density, ρ_0 the medium density, λ the specific volume conductivity, and η the viscosity.

Smoluchowski developed the equation under five assumptions:

- Particles are spherical, nonconducting, and monodispersed.
- Laminar flow around the particles occurs (Reynolds number <1).
- Interparticle interactions are negligible.
- Surface conduction is negligible.
- The double-layer thickness $1/\kappa$ is small compared to the particle radius a ($\kappa a >> 1$).

$$\sigma^{\infty} = \frac{e^2}{k_B T}\sum z_i^2 D_i n_{i\infty}$$

Where D_i is the diffusion coefficient of the *ith* solute species, and $n_{i\infty}$ is the number concentration of electrolyte solution. An improved design cell was developed to determine sedimentation potential, specific conductivity, volume fraction of the solids as well as pH. Two pairs of electrodes are used in this set up, one to measure potential difference and the other for resistance. A flip switch is utilized to avoid polarization of the resistance electrodes and buildup of charge by alternating the current. The pH of the system could be monitored and the electrolyte was drawn into the tube using a vacuum pump.

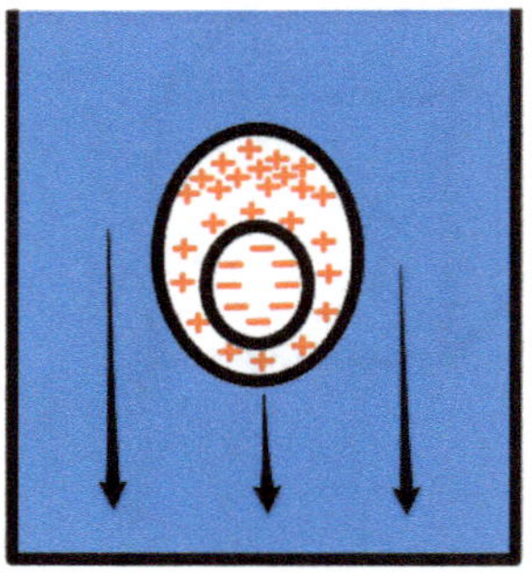

Sedimentation Of A Single Particle Generates a Potential.

Ohshima's model was developed in 1984 and was originally used to analyze the sedimentation velocity of a single charged sphere and the sedimentation potential of a dilute suspension. The model provided below holds true for dilute suspensions of low zeta potential, *i.e.* $e\zeta/\kappa_B T \leq 2$

$$E_s = -\frac{\varepsilon\zeta(\rho - \rho_0)\phi_p}{\sigma^{\infty}\eta} gH(\kappa\alpha) + \vartheta(\zeta^2)$$

Testing

Measuring Sedementation Potential

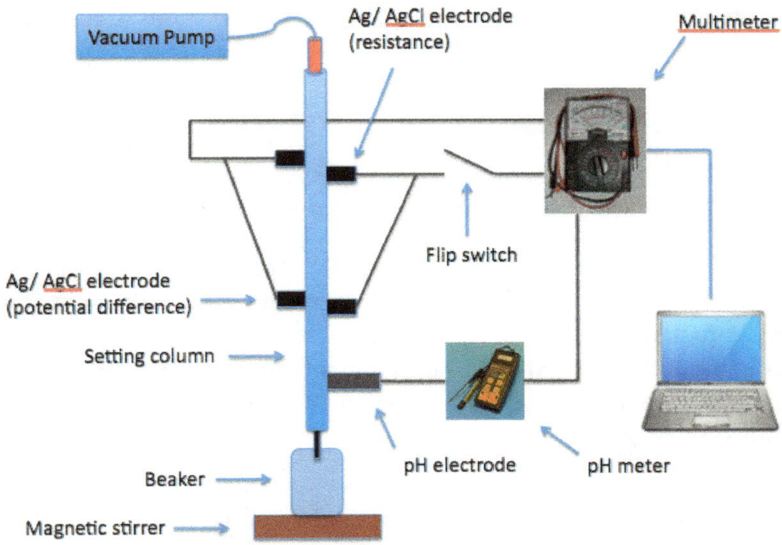

Instrumental Setup to measure Sedimentation Potential.

Sedimentation potential is measured by attaching electrodes to a glass column filled with the dispersion of interest. A voltmeter is attached to measure the potential generated from the suspension. To account for different geometries of the electrode, the column is typically rotated 180 degrees while measuring the potential. This difference in potential through rotation by 180 degrees is twice the sedimentation potential. The zeta potential can be determined through measurement by sedimentation potential, as the concentration, conductivity of the suspension, density of the particle, and potential difference are known. By rotating the column 180 degrees, drift and geometry differences of the column can be ignored.

$$\zeta = \frac{\eta\lambda E_s}{\varepsilon_r\varepsilon_0(\rho - \rho_0)g}$$

When dealing with the case of concentrated systems, the zeta potential can be determined through measurement of the sedimentation potential E_s, from the potential difference relative to the distance between the electrodes. The other parameters represent the following: η the viscosity of the medium; λ the bulk conductivity; ε_r the relative permitivity of the medium; ε_0 the permitivity of free space; ρ the density of the particle; ρ_0 the density of the medium; g is the acceleration due to gravity; and σ^{∞} is the electrical conductivity of the bulk electrolyte solution.

An improved design cell was developed to determine sedimentation potential, specific conductivity, volume fraction of the solids as well as pH. Two pairs of electrodes are used in this set up, one to measure potential difference and the other for resistance. A flip switch is utilized to avoid polarization of the resistance electrodes and buildup of charge by alternating the current. The pH of the system could be monitored and the electrolyte was drawn into the tube using a vacuum pump.

Real World Applications

Applications of Sedimentation Field Flow Fractionation (SFFF)

Sedimentation field flow fractionation (SFFF) is a non-destructive separation technique which can be used for both separation, and collecting fractions. Some applications of SFFF include characterization of particle size of latex materials for adhesives, coatings and paints, colloidal silica for binders, coatings and compounding agents, titanium oxide pigments for paints, paper and textiles, emulsion for soft drinks, and biological materials like viruses and liposomes.

Some main aspects of SFFF include: it provides high-resolution possibilities for size distribution measurements with high precision, the resolution is dependent on experimental conditions, the typical analysis time is 1 to 2 hours, and it is a non-destructive technique which offers the possibility of collecting fraction.

Particle Size Analysis by Sedimentation Field Flow Fractionation

As sedimentation field flow fractionation (SFFF) is one of field flow fractionation separation techniques, it is appropriate for fractionation and characterization of particulate materials and soluble samples in the colloid size range. Differences in interaction between a centrifugal force field and particles with different masses or sizes lead to the separation. An exponential distribution of particles of a certain size or weight is results due to the Brownian motion. Some of the assumptions to develop the theoretical equations include that there is no interaction between individual particles and equilibrium can occur anywhere in separation channels.

Streaming Current

A streaming current and streaming potential are two interrelated electrokinetic phenomena studied in the areas of surface chemistry and electrochemistry. They are an electric current or potential which originates when an electrolyte is driven by a pressure gradient through a channel or porous plug with charged walls.

The first observation of the streaming potential is generally attributed to the German physicist Georg Hermann Quincke in 1859.

Applications

Streaming currents in well-defined geometries are a sensitive method to characterize the zeta potential of surfaces, which is important in the fields of colloid and interface science. In geology, measurements of related spontaneous potential are used for evaluations of formations. Streaming potential has to be considered in design for flow of poorly conductive fluids (e.g., gasoline lines) because of the danger of buildup of high voltages. The streaming current monitor (SCM) is a fundamental tool for monitoring coagulation in wastewater treatment plants. The degree of coagulation of raw water may be monitored by the use of an SCM to provide a positive feedback control of coagulant injection. As the streaming current of the wastewater increases, more coagulant agent is injected into the stream. The higher levels of coagulant agent cause the small colloidal particles to coagulate and sediment out of the stream. Since less colloid particles are in the wastewater stream, the streaming potential decreases. The SCM recognizes this and subsequently reduces the amount of coagulant agent injected into the wastewater stream. The implementation of SCM feedback control has led to a significant materials cost reduction, one that was not realized until the early 1980s. In addition to monitoring capabilities, the streaming current could, in theory, generate usable electrical power. This process, however, has yet to be applied as typical streaming potential mechanical to electrical efficiencies are around 1%.

Origin of the Streaming Current

Adjacent to the channel walls, the charge-neutrality of the liquid is violated due to the presence of the electrical double layer: a thin layer of counterions attracted by the charged surface.

The transport of counterions along with the pressure-driven fluid flow gives rise to a net charge transport: the streaming current. The reverse effect, generating a fluid flow by applying a potential difference, is called electroosmotic flow.

Measurement Method

A typical setup to measure streaming currents consists of two reversible electrodes placed on either side of a fluidic geometry across which a known pressure difference is applied. When both electrodes are held at the same potential, the streaming current is measured directly as the electric current flowing through the electrodes. Alternatively, the electrodes can be left floating, allowing a streaming potential to build up between the two ends of the channel.

A streaming potential is defined as positive when the electric potential is higher on the high pressure end of the flow system than on the low pressure end.

The value of streaming current observed in a capillary is usually related to the zeta potential through the relation:

$$I_{str} = -\frac{\epsilon_{rs}\epsilon_0 a^2 \pi}{\eta} \frac{\Delta P}{L} \zeta.$$

The conduction current, which is equal in magnitude to the streaming current at steady state, is:

$$I_c = K_L a^2 \pi \frac{U_{str}}{L}$$

At steady state, the streaming potential built up across the flow system is given by:

$$U_{str} = \frac{\epsilon_{rs}\epsilon_0 \zeta}{\eta K_L} \Delta P$$

Symbols:

- I_{str} - streaming current under short-circuit conditions, A
- U_{str} - streaming potential at zero net current conditions, V
- I_c - conduction current, A
- ε_{rs} - relative permittivity of the liquid, dimensionless
- ε_0 - electrical permittivity of vacuum, $F{\cdot}m^{-1}$
- η - dynamic viscosity of the liquid, $kg{\cdot}m^{-1}{\cdot}s^{-1}$
- ζ - zeta potential, V
- ΔP - pressure difference, Pa
- L - capillary length, m
- a - capillary radius, m
- K_L - specific conductivity of the bulk liquid, $S{\cdot}m^{-1}$

The equation above is usually referred to as the Helmholtz-Smoluchowski equation.

The above equations assume that:

- the double layer is not too large compared to the pores or capillaries (i.e., $\kappa a \gg 1$), where κ is the reciprocal of the Debye length
- there is no surface conduction (which typically may become important when the zeta potential is large, e.g., $|\zeta| > 50$ mV)
- there is no electrical double layer polarization
- the surface is homogeneous in properties
- there is no axial concentration gradient

- the geometry is that of a capillary/tube.

Colloid Vibration Current

Colloid vibration current is an electroacoustic phenomenon that arises when ultrasound propagates through a fluid that contains ions and either solid particles or emulsion droplets.

The pressure gradient in an ultrasonic wave moves particles relative to the fluid. This motion disturbs the double layer that exists at the particle-fluid interface. The picture illustrates the mechanism of this distortion. Practically all particles in fluids carry a surface charge. This surface charge is screened with an equally charged diffuse layer; this structure is called the double layer. Ions of the diffuse layer are located in the fluid and can move with the fluid. Fluid motion relative to the particle drags these diffuse ions in the direction of one or the other of the particle's poles. The picture shows ions dragged towards the left hand pole. As a result of this drag, there is an excess of negative ions in the vicinity of the left hand pole and an excess of positive surface charge at the right hand pole. As a result of this charge excess, particles gain a dipole moment. These dipole moments generate an electric field that in turn generates measurable electric current. This phenomenon is widely used for measuring zeta potential in concentrated colloids.

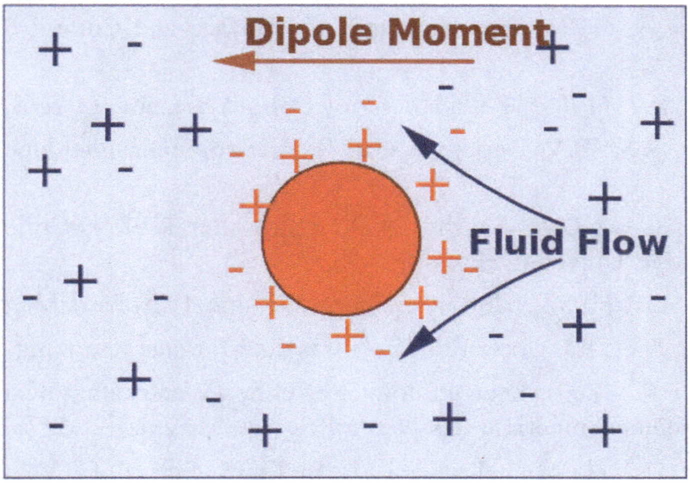

Electric Sonic Amplitude

Electric sonic amplitude is an electroacoustic phenomenon that is the reverse to colloid vibration current. It occurs in colloids, emulsions and other heterogeneousfluids under the influence of an oscillating electric field. This field moves particles relative to the liquid, which generates ultrasound.

Electric sonic amplitude was experimentally discovered by Oja and co-authors in early 1980s. It is also widely used for characterizing zeta potential in dispersions and emulsions. There is review of the theory of this effect, its experimental verification, and multiple applications published by Hunter.

REFERENCES

1. Lyklema, J. (1995) Fundamentals of Interface and Colloid Science, Vol. 2, p. 3.208.
2. Hunter, R.J. (1989) Foundations of Colloid Science, Oxford University Press.
3. Dukhin, S.S. and Derjaguin, B.V. (1974) Electrokinetic Phenomena, J. Willey and Sons.
4. Russel, W.B., Saville, D.A., and Schowalter, W.R. (1989) Colloidal Dispersions, Cambridge University Press.
5. Kruyt, H.R. (1952) Colloid Science, Elsevier. Volume 1, Irreversible systems.
6. Dukhin, A.S. and Goetz, P.J. (2002) Ultrasound for characterizing colloids. Elsevier.
7. Kirby, B.J. (2010). Micro- and Nanoscale Fluid Mechanics: Transport in Microfluidic Devices. Cambridge University Press. ISBN 978-0-521-11903-0.
8. Anand Plappally, Alfred Soboyejo, Norman Fausey, Winston Soboyejo and Larry Brown,"Stochastic Modeling of Filtrate Alkalinity in Water Filtration Devices: Transport through Micro/Nano Porous Clay Based Ceramic Materials" J Nat Env Sci 2010 1(2):96-105.
9. Lyklema, J. (1995). Fundamentals of Interface and Colloid Science. vol. 2. p. 3.208.
10. Hunter, R.J. (1989). Foundations of Colloid Science. Oxford University Press.
11. Dukhin, S.S.; B.V. Derjaguin (1974). Electrokinetic Phenomena. J. Willey and Sons.
12. Russel, W.B.; D.A. Saville; W.R. Schowalter (1989). Colloidal Dispersions. Cambridge University Press.
13. Kruyt, H.R. (1952). Colloid Science. Volume 1, Irreversible systems. Elsevier.
14. Dukhin, A.S.; P.J. Goetz (2002). Ultrasound for characterizing colloids. Elsevier.
15. Reuss, F.F. (1809). "Sur un nouvel effet de l'électricité galvanique". Mémoires de la Société Impériale des Naturalistes de Moscou. 2: 327–337.
16. Hanaor, D.A.H.; Michelazzi, M.; Leonelli, C.; Sorrell, C.C. (2012). "The effects of carboxylic acids on the aqueous dispersion and electrophoretic deposition of ZrO2". Journal of the European Ceramic Society. 32 (1): 235–244. doi:10.1016/j.jeurceramsoc.2011.08.015.
17. von Smoluchowski, M. (1903). "Contribution à la théorie de l›endosmose électrique et de quelques phénomènes corrélatifs". Bull. Int. Acad. Sci. Cracovie. 184.
18. Hückel, E. (1924). "Die kataphorese der kugel". Physik. Z. 25: 204.
19. Overbeek, J.Th.G (1943). "Theory of electrophoresis — The relaxation effect". Koll. Bith.: 287.
20. Booth, F. (1948). "Theory of Electrokinetic Effects". Nature. 161 (4081): 83–86. Bibcode:1948Natur.161...83B. doi:10.1038/161083a0. PMID 18898334.

21. Dukhin, S.S.; N.M. Semenikhin (1970). Koll. Zhur. 32: 366. Missing or empty |title= (help)
22. O›Brien, R.W.; L.R. White (1978). "Electrophoretic mobility of a spherical colloidal particle". J. Chem. Soc. Faraday Trans. 2 (74): 1607. doi:10.1039/F29787401607.
23. G. F. Yao, H. A Computational Model for Simulation of Electroosmotic Flow in Microsystems.
24. Bruus, H. (2007). Theoretical Microfluidics. ISBN 0-19-923509-0.
25. Kirby, B. J. (2010). Micro- and Nanoscale Fluid Mechanics: Transport in Microfluidic Devices: Chapter 6: Electroosmosis. Cambridge University Press.
26. Wise, D. L. and Trantolo, D. J., eds. Remediation of Hazardous Waste Contaminated Soils.
27. Skoog. Principles of Instrumental Analysis. ISBN 0-495-12570-9.
28. Ducree, Jen. myFluidix.com.
29. Ottosen, Lisbeth; Anne J. Pedersen; Inge Rorig-Dalgaard. Salt related problems in brick masonry and electrokinetic removal of salts. Journal of Building Appraisal.
30. "Electro Osmosis Damp Proofing systems – fraud, or the perfect solution to damp – you decide!".
31. Clegg, C. J., Mackean, D. G. (2006) "Advanced Biology – principles & applications" Hodder Stoughton Publhishers, pp. 340–343.
32. Polevoi, V. V. (2003). "Electroosmotic Phenomena in Plant Tissues". Biology Bulletin. 30 (2): 133. doi:10.1023/A:1023285121361.
33. Erlandsson, P. G.; Robinson, N. D. (2011). "Electrolysis-reducing electrodes for electrokinetic devices". Electrophoresis. 32 (6–7): 784–790. doi:10.1002/elps.201000617. PMID 21425174.
34. Derjaguin, B.V., Sidorenko, G.P., Zubashenko, E.A. and Kiseleva, E.B. , Kolloid Zh., vol.9, #5, 335–348 (1947).
35. Dukhin, S.S. & Derjaguin, B.V. "Electrokinetic Phenomena", J. Willey and Sons, 1974.
36. Lyklema, J. (1995). Fundamentals of Interface and Colloid Science. Academic Press.
37. Li, D. (2004). Electrokinetics in Microfluidics. Academic Press.
38. Chang, H.C., Yeo, L. (2009). Electrokinetically Driven Microfluidics and Nanofluidics. Cambridge University Press.
39. http://www.waterhouse-bc.ca/Theory%20of%20the%20Streaming%20Current%20Monitor.pdf
40. "Energy from streaming current and potential". Sensors and Actuators B: Chemical. 111-112: 385–389. doi:10.1016/j.snb.2005.03.039.
41. Kirby, B.J. (2010). Micro- and Nanoscale Fluid Mechanics: Transport in

Microfluidic Devices. Cambridge University Press. ISBN 978-0-521-11903-0.

42. Bruus, H. (2007). Theoretical Microfluidics. Oxford University Press.
43. Karniadakis, G.M., Beskok, A., Aluru, N. (2005). Microflows and Nanoflows. Springer Verlag.
44. "Measurement and Interpretation of Electrokinetic Phenomena", International Union of Pure and Applied Chemistry, Technical Report, published in Pure Appl.Chem., vol 77, 10, pp.1753-1805, 2005 (pdf).
45. Menachem Elimelech and Amy E. Childress, "ZETA POTENTIAL OF REVERSE OSMOSIS MEMBRANES: IMPLICACATIONS FOR MEMBRANE PERFORMANCE". U.S. DEPARTMENT OF THE INTERIOR, Bureau of Reclamation, Denver Office. Water Treatment Technology Program Report No. 10. December 1996.
46. Oja, T., Petersen, G., and Cannon, D. "Measurement of Electric-Kinetic Properties of a Solution", U.S. Patent 4,497,208 1985
47. Hunter, R (1998). "Recent developments in the electroacoustic characterisation of colloidal suspensions and emulsions". Colloids and Surfaces A: Physicochemical and Engineering Aspects. 141: 37. doi:10.1016/S0927-7757(98)00202-7.

CHAPTER 4

Electrolytic Conductivity

Chapter Contents

- **Introduction**
- **Units**
- **Measurement**
- **Theory**
- **Applications**
- **Electrolyte**

INTRODUCTION

Conductivity (or specific conductance) of an electrolyte solution is a measure of ohms. its ability to conduct electricity. The SI unit of conductivity is siemens per meter (S/m).

Conductivity measurements are used routinely in many industrial and environmental applications as a fast, inexpensive and reliable way of measuring the ionic content in a solution. For example, the measurement of product conductivity is a typical way to monitor and continuously trend the performance of water purification systems.

In many cases, conductivity is linked directly to the total dissolved solids (T.D.S.). High quality deionized water has a conductivity of about 5.5 μS/m, typical drinking water in the range of 5-50 mS/m, while sea water about 5 S/m (i.e., sea water›s conductivity is one million times higher than that of deionized water).

Conductivity is traditionally determined by connecting the electrolyte in a Wheatstone bridge. Dilute solutions follow Kohlrausch›s Laws of concentration dependence and additivity of ionic contributions. Lars Onsager gave a theoretical explanation of Kohlrausch›s law by extending Debye–Hückel theory.

UNITS

The SI unit of conductivity is S/m and, unless otherwise qualified, it refers to 25 °C. Often encountered in industry is the traditional unit of μS/cm. 10^6 μS/cm = 10^3 mS/cm = 1 S/cm. The values in μS/cm are lower than those in μS/m by a factor of 100 (i.e., 1 μS/cm = 100 μS/m). Sometimes encountered is mho (reciprocal of ohm): 1 mho/m = 1 S/m. Historically, mhos antedate Siemens by many decades; good vacuum-tube testers, for instance, gave transconductance readings in micromhos.

The commonly used standard cell has a width of 1 cm, and thus for very pure water in equilibrium with air would have a resistance of about 10^6 ohm, known as a megohm. Ultra-pure water could achieve 18 megohms or more. Thus in the past, megohm-cm was used, sometimes abbreviated to "megohm". Sometimes, conductivity is given in "microSiemens" (omitting the distance term in the unit). While this is an error, it can often be assumed to be equal to the traditional μS/cm.

The conversion of conductivity to the total dissolved solids depends on the chemical composition of the sample and can vary between 0.54 and 0.96. Typically, the conversion is done assuming that the solid is sodium chloride, i.e., 1 μS/cm is then equivalent to about 0.64 mg of NaCl per kg of water.

Molar conductivity has the SI unit S m^2 mol^{-1}. Older publications use the unit Ω^{-1} cm^2 mol^{-1}.

MEASUREMENT

The electrical conductivity of a solution of an electrolyte is measured by determining the resistance of the solution between two flat or cylindrical electrodes separated by a fixed distance. An alternating voltage is used in order to avoid electrolysis. The resistance is measured by a conductivity meter. Typical frequencies used are in the range 1–3 kHz. The dependence on the frequency is usually small, but may become appreciable at very high frequencies, an effect known as the Debye–Falkenhagen effect.

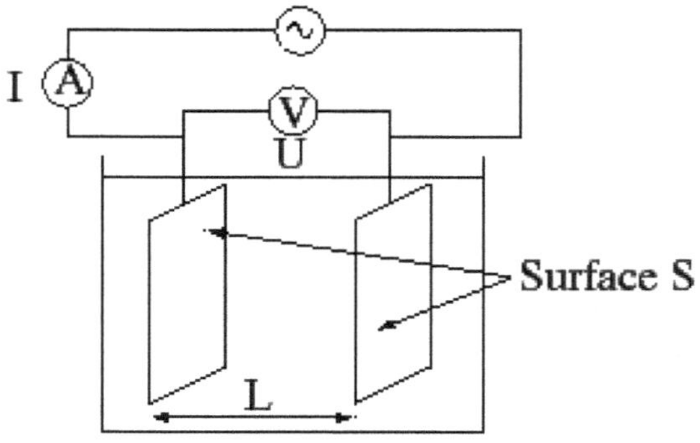

Principle of the measurement.

A wide variety of instrumentation is commercially available. There are two types of cell, the classical type with flat or cylindrical electrodes and a second type based on induction. Many commercial systems offer automatic temperature correction. Tables of reference conductivities are available for many common solutions.

Definitions

Resistance, R, is proportional to the distance, l, between the electrodes and is inversely proportional to the cross-sectional area of the sample, A (noted S on the Figure above). Writing ρ (rho) for the specific resistance (or resistivity),

$$R = \frac{l}{A}\rho.$$

In practice the conductivity cell is calibrated by using solutions of known specific resistance, ρ^*, so the quantities l and A need not be known precisely. If the resistance of the calibration solution is R^*, a cell-constant, C, is derived.

$$R^* = C \times \rho^*$$

The specific conductance, κ (kappa) is the reciprocal of the specific resistance.

$$\kappa = \frac{1}{\rho} = \frac{C}{R}$$

Conductivity is also temperature-dependent. Sometimes the ratio of l and A is called as the cell constant, denoted as G^*, and conductance is denoted as G. Then the specific conductance κ (kappa), can be more conveniently written as

$$\kappa = G^* \times G$$

THEORY

The specific conductance of a solution containing one electrolyte depends on the concentration of the electrolyte. Therefore, it is convenient to divide the specific conductance by concentration. This quotient, termed molar conductivity, is denoted by Λ_m

$$\Lambda_m = \frac{\kappa}{c}$$

Strong Electrolytes

Strong electrolytes are believed to dissociate completely in solution. The conductivity of a solution of a strong electrolyte at low concentration follows Kohlrausch's Law

$$\Lambda_m = \Lambda_m^0 - K\sqrt{c}$$

where Λ_m^0 is known as the limiting molar conductivity, K is an empirical constant and c is the electrolyte concentration. (Limiting here means "at the limit of the infinite dilution".) In effect, the observed conductivity of a strong electrolyte becomes directly proportional to concentration, at sufficiently low concentrations i.e. when

$$\Lambda_m^0 > K\sqrt{c}$$

As the concentration is increased however, the conductivity no longer rises in proportion. Moreover, Kohlrausch also found that the limiting conductivity of an electrolyte;

- λ_+^0 and λ_-^0 are the limiting molar conductivities of the individual ions.

The following table gives values for the limiting molar conductivities for selected ions.

Table of limiting ion conductivity in water at 298°K (approx. 25°C)							
Cations	λ_+^o / mS·m²·mol	Cations	λ_+^o / mS·m²·mol	Anions	λ_-^o / mS·m²·mol	Anions	λ_-^o / mS·m²·mol λ
H^+	34.982	Ba^{2+}	12.728	^-OH	19.8	SO_4^{2-}	15.96
Li^+	3.869	Mg^{2+}	10.612	Cl^-	7.634	$C_2O_4^{2-}$	7.4
Na^+	5.011	La^{3+}	20.88	Br^-	7.84	$HC_2O_4^{1-}$	40.2
K^+	7.352	Rb^+	7.64	I^-	7.68	$HCOO^-$	5.6
NH_4^+	7.34	Cs^+	7.68	NO_3^-	7.144	CO_3^{2-}	7.2
Ag^+	6.192	Be^{2+}	4.50	CH_3COO^-	4.09	HSO_3^{2-}	5.0
Ca^{2+}	11.90			ClO_4^-	6.80	SO_3^{2-}	7.2
$Co(NH_3)_6^{3+}$	10.2			F^-	5.50		

An interpretation of these results was based on the theory of Debye and Hückel, yielding the Debye-Hückel-Onsager theory:

$$\Lambda_m = \Lambda_m^0 - (A + B\Lambda_m^0)\sqrt{c}$$

where *A* and *B* are constants that depend only on known quantities such as temperature, the charges on the ions and the dielectric constant and viscosity of the solvent. As the name suggests, this is an extension of the Debye–Hückel theory, due to Onsager. It is very successful for solutions at low concentration.

Weak Electrolytes

A weak electrolyte is one that is never fully dissociated (i.e. there are a mixture of ions and complete molecules in equilibrium). In this case there is no limit of dilution below which the relationship between conductivity and concentration becomes linear. Instead, the solution becomes ever more fully dissociated at weaker concentrations, and for low concentrations of "well behaved" weak electrolytes, the degree of dissociation of the weak electrolyte becomes proportional to the inverse square root of the concentration.

Typical weak electrolytes are weak acids and weak bases. The concentration of ions in a solution of a weak electrolyte is less than the concentration of the electrolyte itself. For acids and bases the concentrations can be calculated when the value(s) of the acid dissociation constant(s) is(are) known.

For a monoprotic acid, HA, obeying the inverse square root law, with a dissociation constant K_a, an explicit expression for the conductivity as a function of concentration, c, known as Ostwald's dilution law, can be obtained.

$$\frac{1}{\Lambda_m} = \frac{1}{\Lambda_m^0} + \frac{\Lambda_m c}{K_a(\Lambda_m^0)^2}$$

Higher Concentrations

Both Kohlrausch's law and the Debye-Hückel-Onsager equation break down as the concentration of the electrolyte increases above a certain value. The reason for this is that as concentration increases the average distance between cation and anion decreases, so that there is more inter-ionic interaction. Whether this constitutes ion-association is a moot point. However, it has often been assumed that cation and anion interact to form an ion-pair. Thus the electrolyte is treated as if it were like a weak acid and a constant, K, can be derived for the equilibrium

$A^+ + B^- \rightleftharpoons A^+B^-$; $K=[A^+]\ [B^-]/[A^+B^-]$

Davies describes the results of such calculations in great detail, but states that K should not necessarily be thought of as a true equilibrium constant, rather, the inclusion of an "ion-association" term is useful in extending the range of good agreement between theory and experimental conductivity data. Various attempts have been made to extend Onsager's treament to more concentrated solutions.

The existence of a so-called *conductance minimum* has proved to be a controversial subject as regards interpretation. Fuoss and Kraus suggested that it is caused by the formation of ion-triplets, and this suggestion has received some support recently.

Conductivity Versus Temperature

Generally the conductivity of a solution increases with temperature, as the mobility of the ions increases. For comparison purposes reference values are reported at an agreed temperature, usually 298 K (~25 C), although occasionally 20C is used. So called 'compensated' measurements are made at a convenient temperature but the value reported is a calculated value of the expected value of conductivity of the solution, as if it had been measured at the reference temperature. Basic compensation is normally done by assuming a linear increase of conductivity versus temperature of typically 2% per degree Celsius. This value is broadly applicable for most salts at room temperature. Determination of the precise temperature coefficient for a specific solution is simple and instruments are typically capable of applying the derived coefficient (i.e. other than 2%).

Solvent Isotopic Effect

The change in conductivity due to isotopic effect for deuterated electrolytes is sizable.

APPLICATIONS

Notwithstanding the difficulty of theoretical interpretation, measured conductivity is a good indicator of the presence or absence of conductive ions in solution, and measurements are used extensively in many industries. For example, conductivity measurements are used to monitor quality in public water supplies, in hospitals, in boiler water and industries which depend on water quality such as brewing. This type of measurement is not ion-specific; it can sometimes be used to determine the amount of total dissolved solids (T.D.S.) if the composition of the solution and its conductivity behavior are known. It should be noted that conductivity measurements made to determine water purity will not respond to non conductive contaminants (many organic compounds fall into this category), therefore additional purity tests may be required depending on application.

Sometimes, conductivity measurements are linked with other methods to increase the sensitivity of detection of specific types of ions. For example, in the boiler water technology, the boiler blowdown is continuously monitored for "cation conductivity", which is the conductivity of the water after it has been passed through a cation exchange resin. This is a sensitive method of monitoring anion impurities in the boiler water in the presence of excess cations (those of the alkalizing agent usually used for water treatment). The sensitivity of this method relies on the high mobility of H^+ in comparison with the mobility of other cations or anions. Beyond cation conductivity, there are analytical instruments designed to measure Degas conductivity, where conductivity is measured after dissolved carbon dioxide has been removed from the sample, either through reboiling or dynamic degassing.

Conductivity detectors are commonly used with ion chromatography.

ELECTROLYTE

An electrolyte is a substance that produces an electrically conducting solution when dissolved in a polar solvent, such as water. The dissolved electrolyte separates into cations and anions, which disperse uniformly through the solvent. Electrically, such a solution is neutral. If an electric potential is applied to such a solution, the cations of the solution are drawn to the electrode that has an abundance of electrons, while the anions are drawn to the electrode that has a deficit of electrons. The movement of anions and cations in opposite directions within the solution amounts to a current. This includes most soluble salts, acids, and bases. Some gases, such as hydrogen chloride, under conditions of high temperature or low pressure can also function as electrolytes. Electrolyte solutions can also result from the dissolution of some biological (e.g., DNA, polypeptides) and synthetic polymers (e.g., polystyrene sulfonate), termed "polyelectrolytes", which contain charged functional groups. A substance that dissociates into ions in solution acquires the capacity to conduct electricity. Sodium, potassium, chloride, calcium, magnesium, and phosphate are examples of electrolytes, informally known as "lytes".

In medicine, electrolyte replacement is needed when a patient has prolonged vomiting or diarrhea, and as a response to strenuous athletic activity. Commercial electrolyte solutions are available, particularly for sick children (oral rehydration solutions) and athletes (sports drinks). Electrolyte monitoring is important in the treatment of anorexia and bulimia.

Etymology

The word *electrolyte* derives from the Greek *lytós*, meaning “able to be untied or loosened”.

History

Svante Arrhenius put forth, in his 1884 dissertation, his explanation of the fact that solid crystalline salts disassociate into paired charged particles when dissolved, for which he won the 1903 Nobel Prize in Chemistry.

Arrhenius’s explanation was that in forming a solution, the salt dissociates into charged particles, to which Michael Faraday had given the name “ions” many years earlier. Faraday’s belief had been that ions were produced in the process of electrolysis. Arrhenius proposed that, even in the absence of an electric current, solutions of salts contained ions. He thus proposed that chemical reactions in solution were reactions between ions.

Formation

Electrolyte solutions are normally formed when a salt is placed into a solvent such as water and the individual components dissociate due to the thermodynamic interactions between solvent and solute molecules, in a process called “solvation”. For example, when table salt (sodium chloride), NaCl, is placed in water, the salt (a solid) dissolves into its component ions, according to the dissociation reaction

$$NaCl_{(s)} \rightarrow Na^{+}_{(aq)} + Cl^{-}_{(aq)}$$

It is also possible for substances to react with water, producing ions. For example, carbon dioxide gas dissolves in water to produce a solution that contains hydronium, carbonate, and hydrogen carbonate ions.

Molten salts can also be electrolytes as, for example, when sodium chloride is molten, the liquid conducts electricity. In particular, ionic liquids, which are molten salts with melting points below 100 °C, are a type of highly conductive non-aqueous electrolytes and thus have found more and more applications in fuel cells and batteries.

An electrolyte in a solution may be described as “concentrated” if it has a high concentration of ions, or “diluted” if it has a low concentration. If a high proportion of the solute dissociates to form free ions, the electrolyte is strong; if most of the solute does not dissociate, the electrolyte is weak. The properties of electrolytes may be exploited using electrolysis to extract constituent elements and compounds contained within the solution.

Physiological Importance

In physiology, the primary ions of electrolytes are sodium (Na^+), potassium (K^+), calcium (Ca^{2+}), magnesium (Mg^{2+}), chloride (Cl^-), hydrogen phosphate (HPO_4^{2-}), and hydrogen carbonate (HCO_3^-). The electric charge symbols of plus (+) and minus (−) indicate that the substance is ionic in nature and has an imbalanced distribution of electrons, the result of chemical dissociation. Sodium is the main electrolyte found in extracellular fluid and potassium is the main intracellular electrolyte; both are involved in fluid balance and blood pressure control.

All known higher lifeforms require a subtle and complex electrolyte balance between the intracellular and extracellular environments. In particular, the maintenance of precise osmotic gradients of electrolytes is important. Such gradients affect and regulate the hydration of the body as well as blood pH, and are critical for nerve and muscle function. Various mechanisms exist in living species that keep the concentrations of different electrolytes under tight control.

Both muscle tissue and neurons are considered electric tissues of the body. Muscles and neurons are activated by electrolyte activity between the extracellular fluid or interstitial fluid, and intracellular fluid. Electrolytes may enter or leave the cell membrane through specialized protein structures embedded in the plasma membrane called "ion channels". For example, muscle contraction is dependent upon the presence of calcium (Ca^{2+}), sodium (Na^+), and potassium (K^+). Without sufficient levels of these key electrolytes, muscle weakness or severe muscle contractions may occur.

Electrolyte balance is maintained by oral, or in emergencies, intravenous (IV) intake of electrolyte-containing substances, and is regulated by hormones, in general with the kidneys flushing out excess levels. In humans, electrolyte homeostasis is regulated by hormones such as antidiuretic hormones, aldosterone and parathyroid hormones. Serious electrolyte disturbances, such as dehydration and overhydration, may lead to cardiac and neurological complications and, unless they are rapidly resolved, will result in a medical emergency.

Measurement

Measurement of electrolytes is a commonly performed diagnostic procedure, performed via blood testing with ion-selective electrodes or urinalysis by medical technologists. The interpretation of these values is somewhat meaningless without analysis of the clinical history and is often impossible without parallel measurements of renal function. The electrolytes measured most often are sodium and potassium. Chloride levels are rarely measured except for arterial blood gas interpretations, since they are inherently linked to sodium levels. One important test conducted on urine is the specific gravity test to determine the occurrence of an electrolyte imbalance.

Rehydration

In oral rehydration therapy, electrolyte drinks containing sodium and potassium salts replenish the body's water and electrolyte concentrations after dehydration caused by exercise, excessive alcohol consumption, diaphoresis (heavy sweating), diarrhea, vomiting, intoxication or starvation. Athletes exercising in extreme conditions (for three or more hours continuously, e.g. a marathon or triathlon) who do not consume electrolytes risk dehydration (or hyponatremia).

A home-made electrolyte drink can be made by using water, sugar and salt in precise proportions. Pre-made preparations are also available commercially, and also for veterinary use.

Electrolytes are commonly found in fruit juices, sports drinks, milk, nuts, and many fruits and vegetables (whole or in juice form) (e.g., potatoes, avocados).

Electrochemistry

When electrodes are placed in an electrolyte and a voltage is applied, the electrolyte will conduct electricity. Lone electrons normally cannot pass through the electrolyte; instead, a chemical reaction occurs at the cathode, providing electrons to the electrolyte. Another reaction occurs at the anode, consuming electrons from the electrolyte. As a result, a negative charge cloud develops in the electrolyte around the cathode, and a positive charge develops around the anode. The ions in the electrolyte neutralize these charges, enabling the electrons to keep flowing and the reactions to continue.

For example, in a solution of ordinary table salt (sodium chloride, NaCl) in water, the cathode reaction will be

$$2H_2O + 2e^- \rightarrow 2OH^- + H_2$$

and hydrogen gas will bubble up; the anode reaction is

$$2NaCl \rightarrow 2\ Na^+ + Cl_2 + 2e^-$$

and chlorine gas will be liberated. The positively charged sodium ions Na^+ will react toward the cathode, neutralizing the negative charge of OH^- there, and the negatively charged hydroxide ions OH^- will react toward the anode, neutralizing the positive charge of Na^+ there. Without the ions from the electrolyte, the charges around the electrode would slow down continued electron flow; diffusion of H^+ and OH^- through water to the other electrode takes longer than movement of the much more prevalent salt ions. Electrolytes dissociate in water because water molecules are dipoles and the dipoles orient in an energetically favorable manner to solvate the ions.

In other systems, the electrode reactions can involve the metals of the electrodes as well as the ions of the electrolyte.

Electrolytic conductors are used in electronic devices where the chemical reaction at a metal-electrolyte interface yields useful effects.

- In batteries, two materials with different electron affinities are used as electrodes; electrons flow from one electrode to the other outside of the battery, while inside the battery the circuit is closed by the electrolyte's ions. Here, the electrode reactions convert chemical energy to electrical energy.
- In some fuel cells, a solid electrolyte or proton conductor connects the plates electrically while keeping the hydrogen and oxygen fuel gases separated.
- In electroplating tanks, the electrolyte simultaneously deposits metal onto the object to be plated, and electrically connects that object in the circuit.
- In operation-hours gauges, two thin columns of mercury are separated by a small electrolyte-filled gap, and, as charge is passed through the device, the metal dissolves on one side and plates out on the other, causing the visible gap to slowly move along.

- In electrolytic capacitors the chemical effect is used to produce an extremely thin dielectric or insulating coating, while the electrolyte layer behaves as one capacitor plate.
- In some hygrometers the humidity of air is sensed by measuring the conductivity of a nearly dry electrolyte.
- Hot, softened glass is an electrolytic conductor, and some glass manufacturers keep the glass molten by passing a large current through it.

Solid Electrolytes

Solid electrolytes can be mostly divided into four groups:

- Gel electrolytes - closely resemble liquid electrolytes. In essence, they are liquids in a flexible lattice framework. Various additives are often applied to increase the conductivity of such systems.
- Dry polymer electrolytes - differ from liquid and gel electrolytes in the sense that salt is dissolved directly into the solid medium. Usually it is a relatively high dielectric constant polymer (PEO, PMMA, PAN, polyphosphazenes, siloxanes, etc.) and a salt with low lattice energy. In order to increase the mechanical strength and conductivity of such electrolytes, very often composites are used, and inert ceramic phase is introduced. There are two major classes of such electrolytes: polymer-in-ceramic, and ceramic-in-polymer.
- Solid ceramic electrolytes - ions migrate through the ceramic phase by means of vacancies or interstitials within the lattice. There are also glassy-ceramic electrolytes.
- Organic ionic plastic crystals - are a type organic salts exhibiting mesophases (i.e. a state of matter intermediate between liquid and solid), in which mobile ions are orientationally or rotationally disordered while their centers are located at the ordered sites in the crystal structure. They have various forms of disorder due to one or more solid–solid phase transitions below the melting point and have therefore plastic properties and good mechanical flexibility as well as improved electrode|electrolyte interfacial contact. In particular, protic organic ionic plastic crystals (POIPCs), which are solid protic organic salts formed by proton transfer from a Brønsted acid to a Brønsted base and in essence are protic ionic liquids in the molten state, have found to be promising solid-state proton conductors for fuel cells. Examples include 1,2,4-triazolium perfluorobutanesulfonate and imidazolium methanesulfonate.

REFERENCES

1. Gray, James R. (2004). "Conductivity Analyzers and Their Application". In Down, R.D; Lehr, J.H. Environmental Instrumentation and Analysis Handbook. Wiley. pp. 491–510. ISBN 978-0-471-46354-2. Retrieved 10 May 2009.
2. "Water Conductivity". Lenntech. Retrieved 5 January 2013.

3. Bockris, J. O'M.; Reddy, A.K.N; Gamboa-Aldeco , M. (1998). Modern Electrochemistry (2nd. ed.). Springer. ISBN 0-306-45555-2. Retrieved 10 May 2009.
4. Marija Bešter-Rogač and Dušan Habe, "Modern Advances in Electrical Conductivity Measurements of Solutions", Acta Chim. Slov. 2006, 53, 391–395 (pdf)
5. Boyes, W. (2002). Instrumentation Reference Book (3rd. ed.). Butterworth-Heinemann. ISBN 0-7506-7123-8. Retrieved 10 May 2009.
6. Gray, p 495
7. "Conductivity ordering guide" (PDF). EXW Foxboro. 3 October 1999. Retrieved 5 January 2013.
8. "ASTM D1125 - 95(2005) Standard Test Methods for Electrical Conductivity and Resistivity of Water". Retrieved 12 May 2009.
9. Adamson, Aurthur W. (1973). Textbook of Physical Chemistry. London: Academic Press inc. p. 512.
10. Wright, M.R. (2007). An Introduction to Aqueous Electrolyte Solutions. Wiley. ISBN 978-0-470-84293-5.
11. Davies, C.W. (1962). Ion Association. London: Butterworths.
12. Miyoshi, K. (1973). "Comparison of the Conductance Equations of Fuoss - Onsager, Fuoss-Hsia and Pitts with the Data of Bis(2, 9-dimethyl-1, 10-phenanthroline)Cu(I) Perchlorate". Bull. Chem. Soc. Japan. 46 (2): 426–430. doi:10.1246/bcsj.46.426.
13. Fuoss, R.M.; Kraus, C.A. (1935). "Properties of Electrolytic Solutions. XV. Thermodynamic Properties of Very Weak Electrolytes". J. Amer. Chem. Soc. 57: 1–4. doi:10.1021/ja01304a001.
14. Weingärtner, H.; Weiss, V.C.; Schröer, W. (2000). "Ion association and electrical conductance minimum in Debye–Hückel-based theories of the hard sphere ionic fluid". J. Chem. Phys. 113 (2): 762–. Bibcode:2000JChPh.113..762W. doi:10.1063/1.481822.
15. Schröer, W.; Weingärtner, H. (2004). "Structure and criticality of ionic fluids". Pure Appl. Chem. 76 (1): 19–27. doi:10.1351/pac200476010019. pdf
16. http://pubs.acs.org/doi/abs/10.1021/ja970118o
17. "Electrolytic conductivity measurement, Theory and practice" (PDF). Aquarius Technologies Pty Ltd.
18. "Detectors for ion-exchange chromatography". Retrieved 17 May 2009.
19. "The Nobel Prize in Chemistry 1903". Retrieved 5 January 2017.
20. Harris, William; Levey, Judith, eds. (1975). The New Columbia Encyclopedia (4th ed.). New York City: Columbia University. p. 155. ISBN 0-231035-721.
21. McHenry, Charles, ed. (1992). The New Encyclopedia Britannica. 1 (15 ed.). Chicago: Encyclopedia Britannica, Inc. p. 587. ISBN 085-229553-7.
22. Cillispie, Charles, ed. (1970). Dictionary of Scientific Biography (1 ed.). New

York City: Charles Scribner's Sons. pp. 296–302. ISBN 0-684101-122.

23. Jiangshui Luo; Jin Hu; Wolfgang Saak; Rüdiger Beckhaus; Gunther Wittstock; Ivo F. J. Vankelecom; Carsten Agert; Olaf Conrad (2011). "Protic ionic liquid and ionic melts prepared from methanesulfonic acid and 1H-1,2,4-triazole as high temperature PEMFC electrolytes". Journal of Materials Chemistry. 21: 10426–10436. doi:10.1039/C0JM04306K.
24. J, Estevez E; Baquero E; Mora-Rodriguez R (2008). "Anaerobic performance when rehydrating with water or commercially available sports drinks during prolonged exercise in the heat". Applied Physiology, Nutrition, and Metabolism. 33 (2): 290–298. doi:10.1139/H07-188. PMID 18347684.
25. "Rehydration drinks". Webmd.com. 2008-04-28. Retrieved 2010-08-20.
26. "Oral Rehydration Salt Suppliers". Rehydrate.org. 2014-10-07. Retrieved 2014-12-04.
27. Kamil Perzyna; Regina Borkowska; Jaroslaw Syzdek; Aldona Zalewska; Wladyslaw Wieczorek (2011). "The effect of additive of Lewis acid type on lithium–gel electrolyte characteristics". Electrochimica Acta. 57: 58–65. doi:10.1016/j.electacta.2011.06.014.
28. Jiangshui Luo; Annemette H. Jensen; Neil R. Brooks; Jeroen Sniekers; Martin Knipper; David Aili; Qingfeng Li; Bram Vanroy; Michael Wübbenhorst; Feng Yan; Luc Van Meervelt; Zhigang Shao; Jianhua Fang; Zheng-Hong Luo; Dirk E. De Vos; Koen Binnemans; Jan Fransaer (2015). "1,2,4-Triazolium perfluorobutanesulfonate as an archetypal pure protic organic ionic plastic crystal electrolyte for all-solid-state fuel cells". Energy & Environmental Science. 8. doi:10.1039/C4EE02280G.
29. "The Roll-to-Roll Battery Revolution". Ev World. Retrieved 2010-08-20.
30. Syzdek J, Borkowska R, Perzyna K, Tarascon JM, Wieczorek W (2007). "Novel composite polymeric electrolytes with surface-modified inorganic fillers". Journal of Power Sources. 173 (2): 712–720. doi:10.1016/j.jpowsour.2007.05.061. ISSN 0378-7753.
31. Syzdek J, Armand M, Marcinek M, Zalewska A, Żukowska G, Wieczorek W (2010). "Detailed studies on the fillers modification and their influence on composite, poly(oxyethylene)-based polymeric electrolytes". Electrochimica Acta. 55 (4): 1314–1322. doi:10.1016/j.electacta.2009.04.025. ISSN 0013-4686.
32. Syzdek J, Armand M, Gizowska M, Marcinek M, Sasim E, Szafran M, Wieczorek W (2009). "Ceramic-in-polymer versus polymer-in-ceramic polymeric electrolytes—A novel approach". Journal of Power Sources. 194 (1): 66–72. doi:10.1016/j.jpowsour.2009.01.070. ISSN 0378-7753.
33. Jiangshui Luo; Olaf Conrad; Ivo F. J. Vankelecom (2013). "Imidazolium methanesulfonate as a high temperature proton conductor". Journal of Materials Chemistry A. 1. doi:10.1039/C2TA00713D.

CHAPTER 5

Electroanalytical Methods

Chapter Contents

- **Introduction**
- **Potentiometry**
- **Coulometry**
- **Voltammetry**

INTRODUCTION

Electroanalytical methods are a class of techniques in analytical chemistry which study an analyte by measuring the potential (volts) and/or current (amperes) in an electrochemical cell containing the analyte. These methods can be broken down into several categories depending on which aspects of the cell are controlled and which are measured. The three main categories are potentiometry (the difference in electrode potentials is measured), coulometry (the cell's current is measured over time), and voltammetry (the cell's current is measured while actively altering the cell's potential).

POTENTIOMETRY

Potentiometry passively measures the potential of a solution between two electrodes, affecting the solution very little in the process. One electrode is called the reference electrode and has a constant potential, while the other one is an indicator electrode whose potential changes with the composition of the sample. Therefore, the difference of potential between the two electrodes gives an assessment of the composition of the sample. In fact, since potentiometric measurements in a non-destructive measurement, assuming that the electrode is in equilibrium with the solution we are measuring the potential of the solution. Potentiometry usually uses indicator electrodes made *selectively* sensitive to the ion of interest, such as fluoride in fluoride selective electrodes, so that the potential solely depends on the activity of this ion of interest. The time that takes the electrode to establish equilibrium with the solution will affect the sensitivity or accuracy of the measurement. In aquatic environments Platinum due to its high electron transfer kinetics is often used, although an electrode made from several metals can be used in order to enhance to electron transfer kinetics. The most common potentiometric electrode, by far is the glass-membrane electrode used in a pH meter.

A variant of potentiometry is chronopotentiometry which consists in using a constant current and measurement of potential as a function of time. It has been initiated by Weber

COULOMETRY

Coulometry is the name given to a group of techniques in analytical chemistry that determine the amount of matter transformed during an electrolysis reaction by measuring the amount of electricity (in coulombs) consumed or produced. It is named after Charles-Augustin de Coulomb.

There are two basic categories of coulometric techniques. *Potentiostatic coulometry* involves holding the electric potential constant during the reaction using a potentiostat. The other, called *coulometric titration* or *amperostatic coulometry*, keeps the current (measured in amperes) constant using an amperostat.

Potentiostatic Coulometry

Potentiostatic coulometry is a technique most commonly referred to as "bulk electrolysis". The working electrode is kept at a constant potential and the current that flows through the circuit is measured. This constant potential is applied long enough to fully reduce or oxidize all of the electroactive species in a given solution. As the electroactive molecules are consumed, the current also decreases, approaching zero when the conversion is complete. The sample mass, molecular mass, number of electrons in the electrode reaction, and number of electrons passed during the experiment are all related by Faraday's laws. It follows that, if three of the values are known, then the fourth can be calculated.

Bulk electrolysis is often used to unambiguously assign the number of electrons consumed in a reaction observed through voltammetry. It also has the added benefit of producing a solution of a species (oxidation state) which may not be accessible through chemical routes. This species can then be isolated or further characterized while in solution.

The rate of such reactions is not determined by the concentration of the solution, but rather the mass transfer of the electroactive species in the solution to the electrode surface. Rates will increase when the volume of the solution is decreased, the solution is stirred more rapidly, or the area of the working electrode is increased. Since mass transfer is so important the solution is stirred during a bulk electrolysis. However, this technique is generally not considered a hydrodynamic technique, since a laminar flow of solution against the electrode is neither the objective nor outcome of the stirring.

The extent to which a reaction goes to completion is also related to how much greater the applied potential is than the reduction potential of interest. In the case where multiple reduction potentials are of interest, it is often difficult to set an electrolysis potential a "safe" distance (such as 200 mV) past a redox event. The result is incomplete conversion of the substrate, or else conversion of some of the substrate to the more reduced form. This factor must be considered when analyzing the current passed and when attempting to do further analysis/isolation/experiments with the substrate solution.

An advantage to this kind of analysis over electrogravimetry is that it does not require that the product of the reaction be weighed. This is useful for reactions where the product does not deposit as a solid, such as the determination of the amount of arsenic in a sample from the electrolysis of arsenous acid (H_3AsO_3) to arsenic acid (H_3AsO_4).

Coulometric Titration

Coulometric titrations use a constant current system to accurately quantify the concentration of a species. In this experiment, the applied current is equivalent to a titrant. Current is applied to the unknown solution until all of the unknown species is either oxidized or reduced to a new state, at which point the potential of the working electrode shifts dramatically. This potential shift indicates the endpoint. The magnitude of the current (in amperes) and the duration of the current (seconds) can be used to determine the moles of the unknown species in solution. When the volume of the solution is known, then the molarity of the unknown species can be determined.

Advantages of Coulometric Titration

Coulometric titration has the advantage that constant current sources for the generation of titrants are relatively easy to make. • The electrochemical generation of a titrant is much more sensitive and can be much more accurately controlled than the mechanical addition of titrant using a burette drive. For example, a constant current flow of 10 μA for 100ms is easily generated and corresponds to about 10 micrograms of titrant. • The preparation of standard solutions and titer determination is of course no longer necessary. • Chemical substances that are unstable or difficult to handle because of their high volatility or reactivity in solution can also very easily be used as titrants. Examples are bromine, chlorine, Ti3+, Sn2+, Cr2+, and Karl Fischer reagents (iodine). • Coulometric titration can also be performed under inert atmosphere or be remotely controlled e.g. with radioactive substances.

Applications

Karl Fischer Reaction

The Karl Fischer reaction uses a coulometric titration to determine the amount of water in a sample. It can determine concentrations of water on the order of milligrams per liter. It is used to find the amount of water in substances such as butter, sugar, cheese, paper, and petroleum.

The reaction involves converting solid iodine into hydrogen iodide in the presence of sulfur dioxide and water. Methanol is most often used as the solvent, but ethylene glycol and diethylene glycol also work. Pyridine is often used to prevent the buildup of sulfuric acid, although the use of imidazole and diethanolamine for this role are becoming more common. All reagents must be anhydrous for the analysis to be quantitative. The balanced chemical equation, using methanol and pyridine, is:

$$[C_5H_5NH]SO_3CH_3 + I_2 + H_2O + 2C_5H_5N \longrightarrow [C_5H_5NH]SO_4CH_3 + 2[C_5H_5NH]I$$

In this reaction, a single molecule of water reacts with a molecule of iodine. Since this technique is used to determine the water content of samples, atmospheric humidity could alter the results. Therefore, the system is usually isolated with drying tubes or placed in an inert gas container. In addition, the solvent will undoubtedly have some water in it so the solvent's water content must be measured to compensate for this inaccuracy.

To determine the amount of water in the sample, analysis must first be performed using either back or direct titration. In the direct method, just enough of the reagents will be added to completely use up all of the water. At this point in the titration, the current approaches zero. It is then possible to relate the amount of reagents used to the amount of water in the system via stoichiometry. The back-titration method is similar, but involves the addition of an excess of the reagent. This excess is then consumed by adding a known amount of a standard solution with known water content. The result reflects the water content of the sample and the standard solution. Since the amount of water in the standard solution is known, the difference reflects the water content of the sample.

Determination of Film Thickness

Coulometry can be used in the determination of the thickness of metallic coatings. This is performed by measuring the quantity of electricity needed to dissolve a well-defined area of the coating. The film thickness Δ is proportional to the constant current i, the molecular weight M of the metal, the density p of the metal, and the surface area A:

$$\Delta \propto \frac{iM}{A\rho}$$

The electrodes for this reaction are often platinum electrode and an electrode that relates to the reaction. For tin coating on a copper wire, a tin electrode is used, while a sodium chloride-zinc sulfate electrode would be used to determine the zinc film on a piece of steel. Special cells have been created to adhere to the surface of the metal to measure its thickness. These are basically columns with the internal electrodes with magnets or weights to attach to the surface. The results obtained by this coulometric method are similar to those achieved by other chemical and metallurgic techniques.

Coulometers

Electronic Coulometer

The electronic coulometer is based on the application of the operational amplifier in the "integrator"-type circuit. The current passed through the resistor R1 makes a potential drop which is integrated by operational amplifier on the capacitor plates; the higher current, the larger the potential drop. The current need not be constant. In such scheme V_{out} is proportional of the passed charge. Sensitivity of the coulometer can be changed by choosing of the appropriate value of R_1.

Electrochemical Coulometers

There are three common types of coulometers based on electrochemical processes:

- Copper coulometer
- Mercury coulometer
- Hofmann voltameter

"Voltameter" is a synonym for "coulometer".

Copper Coulometer

The copper coulometer is a one application for the copper-copper(II) sulfate electrode. Such a coulometer consists of two identical copper electrodes immersed in slightly acidic pH-buffered solution of copper(II) sulfate. Passing of current through the element leads to the anodic dissolution of the metal on anode and simultaneous deposition of copper ions on the cathode. These reactions have 100% efficiency over a wide range of current density.

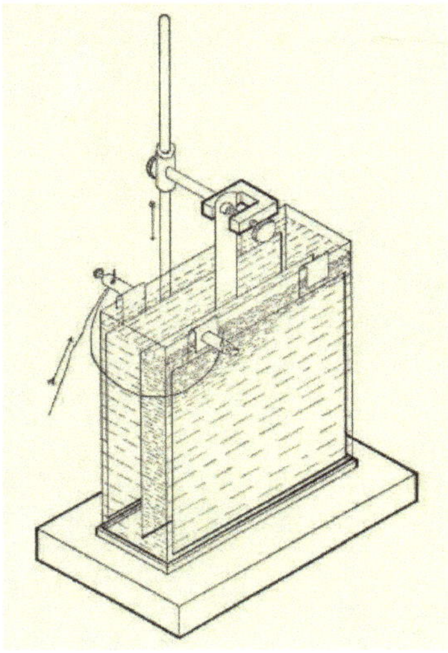

Copper coulometer.

Calculation

The amount of electric charge (quantity of electricity) passed through the cell can easily be determined by measuring the change in mass of either electrode and calculating:

$$Q = \frac{z_{Cu} \Delta m F}{M_{Cu}},$$

where:

- Q is the quantity of electricity (coulombs)
- Δm is the mass transported (gm)
- z_{Cu} is the charge of the copper ions, equal to +2
- F is the Faraday constant (96485.3383 coulombs per mole)
- M_{Cu} is the atomic weight of copper, equal to 63.546 grams per mole.

Although this apparatus is interesting from a theoretical and historical point of view, present-day electronic measurement of time and electric current provide in their multiplication the amount of passed coulombs much easier, with greater precision, and in a shorter period of time than is possible by weighing the electrodes.

Mercury Coulometer

Mercury coulometer is an electroanalytical chemistry device using mercury to determine the amount of matter transformed (in coulombs) during the following reaction:

$Hg^{2+} + 2e^- = Hg^o$

These oxidation/reduction processes have 100% efficiency with the wide range of the current densities. Measuring of the quantity of electricity (coulombs) is based on the changes of the mass of the mercury electrode. Mass of the electrode can be increased during cathodic deposition of the mercury ions or decreased during the anodic dissolution of the metal.

$$Q = \frac{2 \Delta m F}{M_{\mathrm{Hg}}},$$

where

Q, quantity of electricity;

Δm, mass changes;

F, Faraday constant

M_{Hg}, the molar mass of mercury

Construction

This coulometer has different constructions but all of them are based on mass measurements. One of the most significant construction is shown on the picture. It consists of two reservoirs connected by a thin graduated capillary tube. All system contains solution of the mercury(II)-ions solution. Each of the reservoirs have an electrode immersed into a drop of mercury. Another small drop of mercury is inserted into the capillary. When the current is turned on, it will initiate dissolution of the metallic mercury on the one side of the drop in the capillary and deposition on the other side of the same drop. This drop starts to move. Because of the 100% efficiency of the deposition/dissolution of the mercury under the current influence, mass or volume of this small drop will be a constant and its movement will be linearly correlated with the passed charge. If the direction of the current is changed, the drop starts move in the opposite direction. Sensitivity of this type of coulometers depends on the diameter of the capillary.

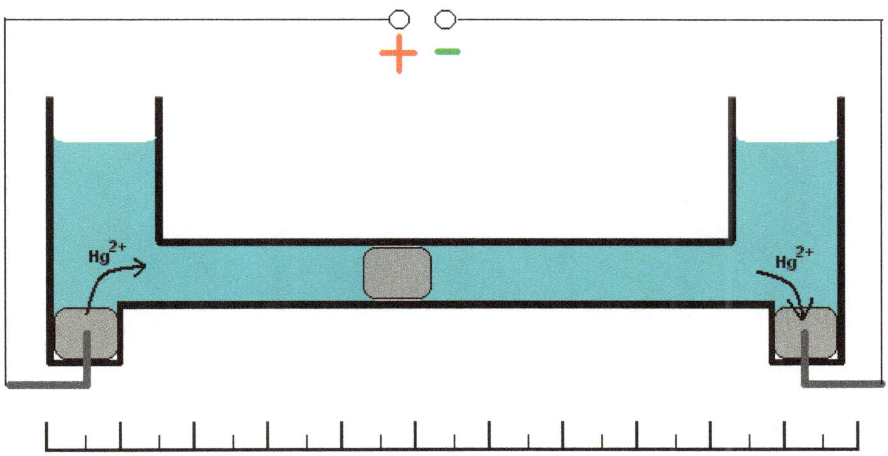

Principle scheme of the mercury coulometer.

Hofmann Voltameter

A Hofmann voltameter is an apparatus for electrolyzing water, invented by August Wilhelm von Hofmann (1818–1892) in 1866. It consists of three joined upright cylinders, usually glass. The inner cylinder is open at the top to allow addition of water and an ionic compound to improve conductivity, such as a small amount of sulphuric acid. A platinum electrode is placed inside the bottom of each of the two side cylinders, connected to the positive and negative terminals of a source of electricity. When current is run through Hofmann's voltameter, gaseous oxygen forms at the anode and gaseous hydrogen at the cathode. Each gas displaces water and collects at the top of the two outer tubes.

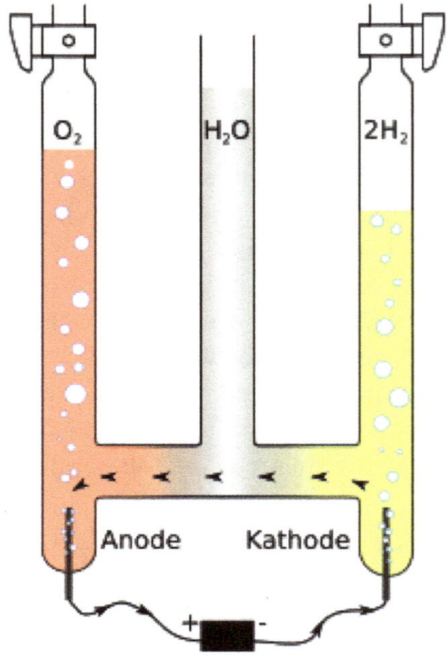

Hofmann voltameter.

Name

The name 'voltameter' was coined by Daniell, who shortened Faraday's original name of "volta-electrometer". Hofmann voltameters are no longer used as electrical measuring devices. However, before the invention of the ammeter, voltameters were often used to measure direct current, since current through a voltameter with iron or copper electrodes electroplates the cathode with an amount of metal from the anode directly proportional to the total coulombs of charge transferred (Faraday's law of electrolysis). The modern name is "electrochemical coulometer". Although the correct spelling of Hofmann contains only one 'f', it is often incorrectly depicted as Hoffmann.

Uses

The amount of electricity that has passed through the system can then be determined by weighing the cathode. Thomas Edison used voltameters as electricity meters. (A Hofmann voltameter cannot be used to weigh electric current in this fashion, as the platinum electrodes are too inert for plating.)

A Hofmann voltameter is often used as a demonstration of stoichiometric principles, as the two-to-one ratio of the volumes of hydrogen and oxygen gas produced by the apparatus illustrates the chemical formula of water, H_2O. However, this is only true if oxygen and hydrogen gases are assumed to be diatomic. If hydrogen gas were monatomic and oxygen diatomic, the gas volume ratio would be 4:1. The volumetric composition of water is the ratio by volume of hydrogen to oxygen present. This value is 2:1 experimentally; this value is determined using Hofmann's water voltameter.

VOLTAMMETRY

Voltammetry is a category of electroanalytical methods used in analytical chemistry and various industrial processes. In voltammetry, information about an analyte is obtained by measuring the current as the potential is varied.

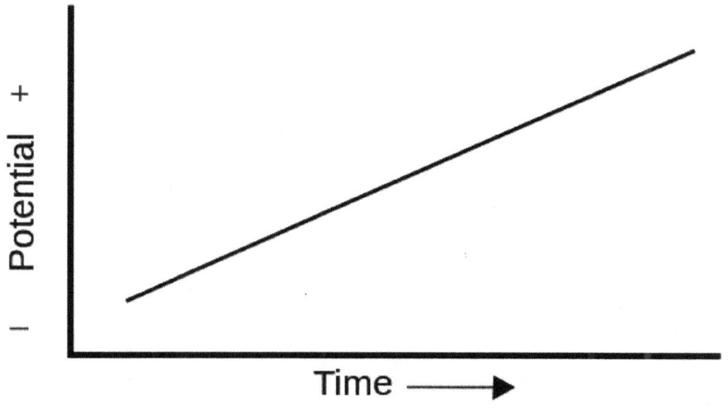

Linear potential sweep.

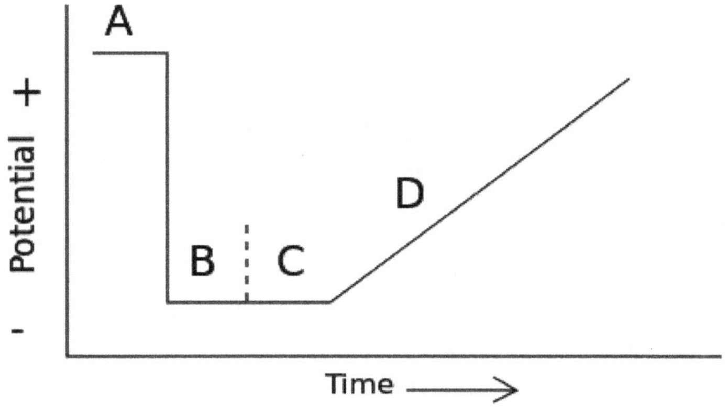

Potential as a function of time for anodic stripping voltammetry.

Three Electrode System

Voltammetry experiments investigate the half-cell reactivity of an analyte. Voltammetry is the study of current as a function of applied potential. These curves I = f(E) are called voltammograms. The potential is varied arbitrarily either step by step or continuously, and the actual current value is measured as the dependent variable. The opposite, i.e., amperometry, is also possible but not common. The shape of the curves depends on the speed of potential variation (nature of driving force) and on whether the solution is stirred or quiescent (mass transfer). Most experiments control the potential (volts) of an electrode in contact with the analyte while measuring the resulting current (amperes).

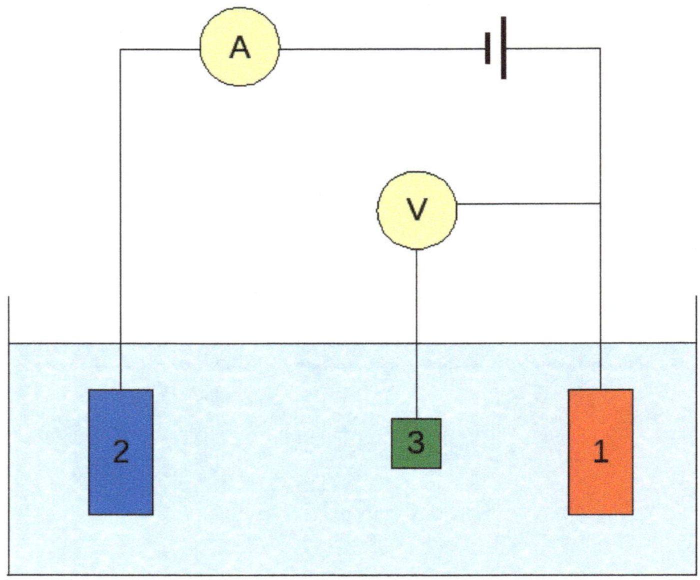

Three-electrode setup: (1) working electrode; (2) counter electrode; (3) reference electrode.

To conduct such an experiment one requires at least two electrodes. The working electrode, which makes contact with the analyte, must apply the desired potential in a controlled way and facilitate the transfer of charge to and from the analyte. A second electrode acts as the other half of the cell. This second electrode must have a known potential with which to gauge the potential of the working electrode, furthermore it must balance the charge added or removed by the working electrode. While this is a viable setup, it has a number of shortcomings. Most significantly, it is extremely difficult for an electrode to maintain a constant potential while passing current to counter redox events at the working electrode.

To solve this problem, the roles of supplying electrons and providing a reference potential are divided between two separate electrodes. The reference electrode is a half cell with a known reduction potential. Its only role is to act as reference in measuring and controlling the working electrode's potential and at no point does it pass any current. The auxiliary electrode passes all the current needed to balance the current observed at the working electrode. To achieve this current, the auxiliary will often swing to extreme potentials at the edges of the solvent window, where it oxidizes or reduces the solvent or supporting electrolyte. These electrodes, the working, reference, and auxiliary make up the modern three electrode system.

There are many systems which have more electrodes, but their design principles are generally the same as the three electrode system. For example, the rotating ring-disk electrode has two distinct and separate working electrodes, a disk and a ring, which can be used to scan or hold potentials independently of each other. Both of these electrodes are balanced by a single reference and auxiliary combination for an overall four electrode design. More complicated experiments may add working electrodes as required and at times reference or auxiliary electrodes.

In practice it can be important to have a working electrode with known dimensions and surface characteristics. As a result, it is common to clean and polish working electrodes regularly. The auxiliary electrode can be almost anything as long as it doesn't react with the bulk of the analyte solution and conducts well. It is (or was?) common to use mercury as working electrode e.g. DME and HMDE, and also as auxiliary, and the voltammetry method is then known as polarography. The reference is the most complex of the three electrodes; there are a variety of standards used and it is worth investigating elsewhere. For non-aqueous work, IUPAC recommends the use of the ferrocene/ferrocenium couple as an internal standard. In most voltammetry experiments, a bulk electrolyte (also known as a supporting electrolyte) is used to minimize solution resistance. It is possible to run an experiment without a bulk electrolyte, but the added resistance greatly reduces the accuracy of the results. With room temperature ionic liquids, the solvent can act as the electrolyte.

Theory

Data analysis requires the consideration of kinetics in addition to thermodynamics, due to the temporal component of voltammetry. Idealized theoretical electrochemical thermodynamic relationships such as the Nernst equation are modeled without a time component. While these models are insufficient alone to describe the dynamic aspects of voltammetry, models like the Tafel equation and Butler-Volmer equation lay the groundwork for the modified voltammetry relationships that relate theory to observed results.

Types of Voltammetry

- Linear sweep voltammetry
- Staircase voltammetry
- Squarewave voltammetry
- Cyclic voltammetry - A voltammetric method that can be used to determine diffusion coefficients and half cell reduction potentials.
- Anodic stripping voltammetry - A quantitative, analytical method for trace analysis of metal cations. The analyte is deposited (electroplated) onto the working electrode during a deposition step, and then oxidized during the stripping step. The current is measured during the stripping step.
- Cathodic stripping voltammetry - A quantitative, analytical method for trace analysis of anions. A positive potential is applied, oxidizing the mercury electrode

and forming insoluble precipitates of the anions. A negative potential then reduces (strips) the deposited film into solution.

- Adsorptive stripping voltammetry - A quantitative, analytical method for trace analysis. The analyte is deposited simply by adsorption on the electrode surface (i.e., no electrolysis), then electrolyzed to give the analytical signal. Chemically modified electrodes are often used.
- Alternating current voltammetry
- Polarography - a subclass of voltammetry where the working electrode is a dropping mercury electrode (DME), useful for its wide cathodic range and renewable surface.
- Rotated electrode voltammetry - A hydrodynamic technique in which the working electrode, usually a rotating disk electrode (RDE) or rotating ring-disk electrode (RRDE), is rotated at a very high rate. This technique is useful for studying the kinetics and electrochemical reaction mechanism for a half reaction.
- Normal pulse voltammetry
- Differential pulse voltammetry
- Chronoamperometry

Linear Sweep Voltammetry

Linear sweep voltammetry is a voltammetric method where the current at a working electrode is measured while the potential between the working electrode and a reference electrode is swept linearly in time. Oxidation or reduction of species is registered as a peak or trough in the current signal at the potential at which the species begins to be oxidized or reduced.

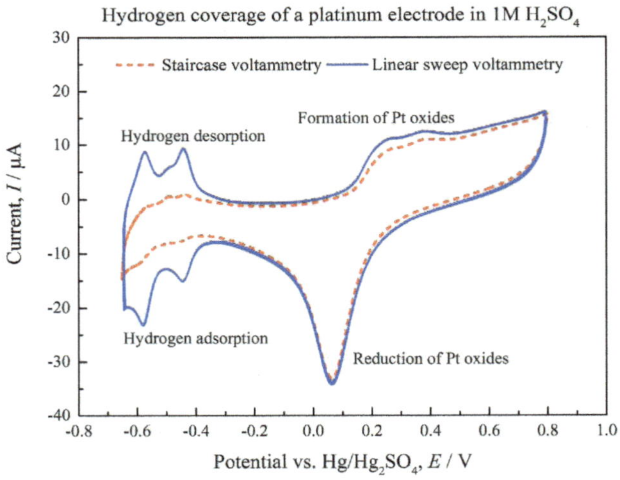

Comparison of the current response of a platinum disc electrode in 1 M sulphuric acid given by linear sweep voltammetry and staircase voltammetry methods.

Experimental Method

The experimental setup for linear sweep voltammetry utilizes a potentiostat and a three-electrode setup to deliver a potential to a solution and monitor its change in current. The three-electrode setup consists of a working electrode, an auxiliary electrode, and a reference electrode. The potentiostat delivers the potentials through the three-electrode setup. A potential, E, is delivered through the working electrode. The slope of the potential vs. time graph is called the scan rate and can range from mV/s to 1,000,000 V/s. The working electrode is one of the electrodes at which the oxidation/reduction reactions occur—the processes that occur at this electrode are the ones being monitored. The auxiliary electrode (or counter electrode) is the one at which a process opposite from the one taking place at the working electrode occurs. The processes at this electrode are not monitored. The equation below gives an example of a reduction occurring at the surface of the working electrode. E_s is the reduction potential of A (if the electrolyte and the electrode are in their standard conditions, then this potential is a standard reduction potential). As E approaches E_s the current on the surface increases and when $E=E_s$ then the concentration of $[A] = [A^-]$ at the surface. As the molecules on the surface of the working electrode are oxidized/reduced they move away from the surface and new molecules come into contact with the surface of the working electrode. The flow of electrons into or out of the electrode causes the current. The current is a direct measure of the rate at which electrons are being exchanged through the electrode-electrolyte interface. When this rate becomes higher than the rate at which the oxidizing or reducing species can diffuse from the bulk of the electrolyte to the surface of the electrode, the current reaches a plateau or exhibits a peak.

$$A + e^- = A^-, E_s = 0.00V$$

Reduction of molecule A at the surface of the working electrode.

The auxiliary and reference electrode work in unison to balance out the charge added or removed by the working electrode. The auxiliary electrode balances the working electrode, but in order to know how much potential it has to add or remove it relies on the reference electrode. The reference electrode has a known reduction potential. The auxiliary electrode tries to keep the reference electrode at a certain reduction potential and to do this it has to balance the working electrode.

Characterization

Linear sweep voltammetry can identify unknown species and determine the concentration of solutions. E1/2 can be used to identify the unknown species while the height of the limiting current can determine the concentration. The sensitivity of current changes vs. voltage can be increased by increasing the scan rate. Higher potentials per second result in more oxidation/reduction of a species at the surface of the working electrode.

Variations

For reversible reactions cyclic voltammetry can be used to find information about the forward reaction and the reverse reaction. Like linear sweep voltammetry, cyclic voltammetry applies a linear potential over time and at a certain potential the potentiostat will reverse the potential applied and sweep back to the beginning point. Cyclic voltammetry provides information about the oxidation and reduction reactions.

Applications

While cyclic voltammetry is applicable to most cases where linear sweep voltammetry is used, there are some instances where linear sweep voltammetry is more useful. In cases where the reaction is irreversible cyclic voltammetry will not give any additional data that linear sweep voltammetry would give us. In one example, linear voltammetry was used to examine direct methane production via a biocathode. Since the production of methane from CO2 is an irreversible reaction, cyclic voltammetry did not present any distinct advantage over linear sweep voltammetry. This group found that the biocathode produced higher current densities than a plain carbon cathode and that methane can be produced from a direct electric current without the need of hydrogen gas.

Staircase Voltammetry

Staircase voltammetry is a derivative of linear sweep voltammetry. In linear sweep voltammetry the current at a working electrode is measured while the potential between the working electrode and a reference electrode is swept linearly in time. Oxidation or reduction of species is registered as a peak or trough in the current signal at the potential at which the species begins to be oxidized or reduced. In staircase voltammetry the potential sweep is a series of stair steps. The current is measured at the end of each potential change, right before the next, so that the contribution to the current signal from the capacitive charging current is reduced.

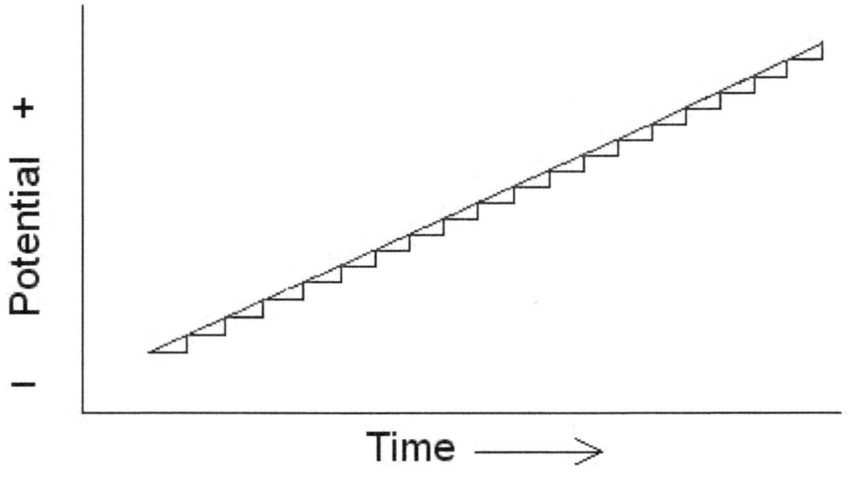

Staircase potential sweep (black) set against a linear potential sweep (blue).

Squarewave Voltammetry

Squarewave voltammetry (SWV) is a form of linear potential sweep voltammetry which has found numerous applications in various fields, including within medicinal and various sensing communities.

When first reported by Barker in 1957, the working electrode utilized was primarily a dropping mercury electrode (DME). When using a DME, the surface area of the mercury drop is constantly changing throughout the course of the experiment; for this reason, complex mathematical modeling was at times required in order to analyze collected electrochemical data. The squarewave voltammetric technique allowed for the collection of the desired electrochemical data within one mercury drop, meaning that the need for mathematical modeling to account for the changing working electrode surface area was no longer needed. In short, the introduction and development of this technique allowed for the rapid collection of reliable and easily reproducible electrochemical data using DME or SDME working electrodes. With continued improvements from many electrochemists (particularly the Osteryoungs), SWV is now one of the primary voltammetric techniques available on modern potentiostats.

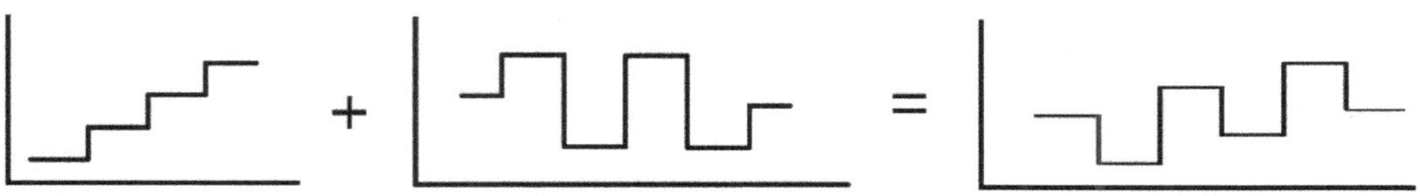

Image explaining the Origins of the Potential Waveform in Squarewave Voltammetric Analyses.

Theory of SWV

In a squarewave voltammetric experiment, the current at a (usually stationary) working electrode is measured while the potential between the working electrode and a reference electrode is swept linearly in time. The potential waveform can be viewed as a superposition of a regular squarewave onto an underlying staircase (see figure above); in this sense, SWV can be considered a modification of staircase voltammetry.

The current is sampled at two times - once at the end of the forward potential pulse and again at the end of the reverse potential pulse (in both cases immediately before the potential direction is reversed). As a result of this current sampling technique, the contribution to the current signal resulting from capacitive (sometimes referred to as non-faradaic or charging) current is minimal. As a result of having current sampling at two different instances per squarewave cycle, two current waveforms are collected - both have diagnostic value, and are therefore preserved. When viewed in isolation, the forward and reverse current waveforms mimic the appearance of a cyclic voltammogram (which corresponds to the anodic or cathodic halves, however, is dependent upon experimental conditions).

Despite both the forward and reverse current waveforms having diagnostic worth, it is almost always the case in SWV for the potentiostat software to plot a differential current waveform derived by subtracting the reverse current waveform from the forward current waveform. This differential curve is then plotted against the applied potential. Peaks in the differential current vs. applied potential plot are indicative of redox processes, and the magnitudes of the peaks in this plot are proportional to the concentrations of the various redox active species according to:

$$\Delta i_p = \frac{nFAD_0^{1/2}C_0^*}{(\pi t_p)^{1/2}}\Delta\Psi_p$$

where $\mathbf{\Delta i_p}$ is the differential current peak value, A is the surface area of the electrode, C_0^* is the concentration of the specices, D_0 is the diffusivity of the species, t_p is the pulse width, and $\mathbf{\Delta\Psi_p}$ is a dimensionless parameter which gauges the peak height in SWV relative to the limiting response in normal pulse voltammetry.

For more information regarding the theory of SWV, please review this seminal paper.

Renewal of Diffusion Layer

It is important to note that in squarewave volammetric analyses, the diffusion layer is not renewed between potential cycles. Thus, it is not possible/accurate to view each cycle in isolation; the conditions present for each cycle is a complex diffusion layer which has evolved through all prior potential cycles. The conditions for a particular cycle are also a function of electrode kinetics, along with other electrochemical considerations.

Applications

Because of the minimal contributions from non-faradaic currents, the use of a differential current plot instead of separate forward and reverse current plots, and significant time evolution between potential reversal and current sampling, high sensitivity screening can be obtained utilizing SWV. For this reason, squarewave voltammetry has been utilized in numerous electrochemical measurements and can be viewed as an improvement to other electroanalytical techniques. For instance, SWV suppressed background currents much more effectively than cyclic voltammetry - for this reason, analyte concentrations on the nanomolar scale can be registered utilizing SWV over CV.

SWV analysis has been used recently in the development of a voltammetric catechol sensor, in the analysis of a large number of pharmaceuticals, and in the development and construction of a 2,4,6-TNT and 2,4-DNT sensor

In addition to being utilized in independent analyses, SWV has also been coupled with other analytical techniques, including but not limited to thin-layer chromatography (TLC) and high-pressure liquid chromatography.

Cyclic Voltammetry

Cyclic voltammetry (CV) is a type of potentiodynamic electrochemical measurement. In a cyclic voltammetry experiment, the working electrode potential is ramped linearly versus time. Unlike in linear sweep voltammetry, after the set potential is reached in a CV experiment, the working electrode's potential is ramped in the opposite direction to return to the initial potential. These cycles of ramps in potential may be repeated as many times as needed. The current at the working electrode is plotted versus the applied voltage (that is, the working electrode's potential) to give the cyclic voltammogram trace. Cyclic voltammetry is generally used to study the electrochemical properties of an analyte in solution.

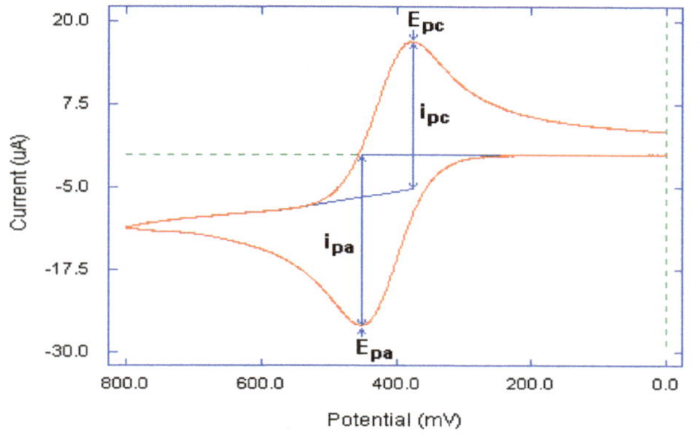

Typical cyclic voltammogram where i_{pc} and i_{pa} show the peak cathodic and anodic current respectively for a reversible reaction. The y-axis shows the negative current.

Experimental Method

In cyclic voltammetry, the electrode potential ramps linearly versus time in cyclical phases. The rate of voltage change over time during each of these phases is known as the experiment's scan rate (V/s). The potential is measured between the working electrode and the reference electrode, while the current is measured between the working electrode and the counter electrode. These data are plotted as current (i) versus applied potential (E, often referred to as just 'potential'). In figure, during the initial forward scan (from t_0 to t_1) an increasingly reducing potential is applied; thus the cathodic current will, at least initially, increase over this time period assuming that there are reducible analytes in the system. At some point after the reduction potential of the analyte is reached, the cathodic current will decrease as the concentration of reducible analyte is depleted. If the redox couple is reversible then during the reverse scan (from t_1 to t_2) the reduced analyte will start to be re-oxidized, giving rise to a current of reverse polarity (anodic current) to before. The more reversible the redox couple is, the more similar the oxidation peak will be in shape to the reduction peak. Hence, CV data can provide information about redox potentials and electrochemical reaction rates.

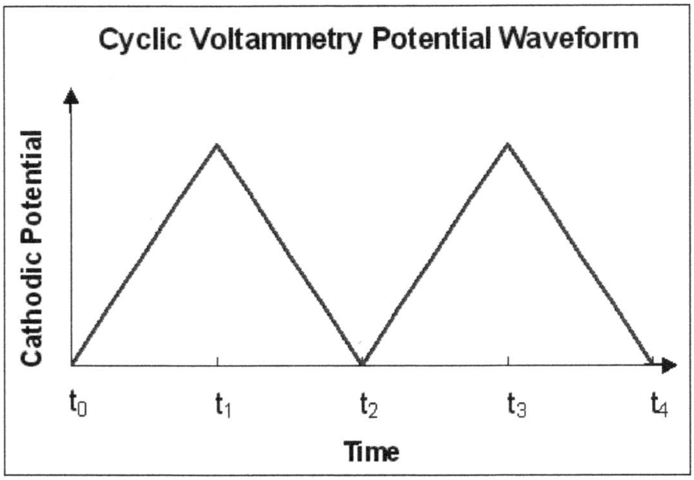

Cyclic voltammetry waveform.

For instance, if the electron transfer at the working electrode surface is fast and the current is limited by the diffusion of analyte species to the electrode surface, then the peak current will be proportional to the square root of the scan rate. This relationship is described by the Cottrell equation. In this situation, the CV experiment only samples a small portion of the solution, i.e., the diffusion layer at the electrode surface.

Characterization

The utility of cyclic voltammetry is highly dependent on the analyte being studied. The analyte has to be redox active within the potential window to be scanned.

Reversible Couples

Often the analyte displays a reversible CV wave, which is observed when all of the initial analyte can be recovered after a forward and reverse scan cycle. Although such reversible couples are simpler to analyze, they contain less information than more complex waveforms.

The waveform of even reversible couples is complex owing to the combined effects of polarization and diffusion. The difference between the two peak potentials (E_p), $\mathbf{\Delta E_p}$, is of particular interest.

$$\Delta E_p = |E_{pc} - E_{pa}|$$

This difference mainly results from the effects of analyte diffusion rates. In the ideal case of a reversible 1e- couple, ΔE_p is 59 mV. Typical values observed experimentally are greater, often approaching 70 or 80 mV. The waveform is also affected by the rate of electron transfer, usually discussed as the activation barrier for electron transfer. A theoretical description of polarization overpotential is in part described by the Butler-Volmer equation and Cottrell equation equations. In an ideal system the relationships reduces to $E_{pa} - E_{pc} = \frac{56.5\ \mathrm{mV}}{n}$ for an *n* electron process.

Focusing on current, reversible couples are characterized by $i_{pa}/i_{pc} = 1$.

When a reversible peak is observed, thermodynamic information in the form of a half cell potential $E^0_{1/2}$ can be determined. When waves are semi-reversible (i_{pa}/i_{pc} is close but not equal to 1), it may be possible to determine even more specific information.

Nonreversible Couples

Many redox processes observed by CV are quasi-reversible or non-reversible. In such cases the thermodynamic potential $E^0_{1/2}$ is often deduced by simulation. The irreversibility is indicated by $i_{pa}/i_{pc} \neq 1$. Deviations from unity are attributable to a subsequent chemical reaction that is triggered by the electron transfer. Such EC processes can be complex, involving isomerization, dissociation, association, etc.

Experimental Setup

CV experiments are conducted on a solution in a cell fitted with electrodes. The solution consists of the solvent, in which is dissolved electrolyte and the species to be studied.

The Cell

A standard CV experiment employs a cell fitted with three electrodes: reference electrode, working electrode, and counter electrode. This combination is sometimes referred to as a three-electrode setup. Electrolyte is usually added to the sample solution to ensure sufficient conductivity. The solvent, electrolyte, and material composition of the working electrode will determine the potential range that can be accessed during the experiment.

The electrodes are immobile and sit in unstirred solutions during cyclic voltammetry. This "still" solution method gives rise to cyclic voltammetry's characteristic diffusion-controlled peaks. This method also allows a portion of the analyte to remain after reduction or oxidation so that it may display further redox activity. Stirring the solution between cyclic voltammetry traces is important in order to supply the electrode surface with fresh analyte for each new experiment. The solubility of an analyte can change drastically with its overall charge; as such it is common for reduced or oxidized analyte species to precipitate out onto the electrode. This layering of analyte can insulate the electrode surface, display its own redox activity in subsequent scans, or otherwise alter the electrode surface in a way that affects the CV measurements. For this reason it is often necessary to clean the electrodes between scans.

Common materials for the working electrode include glassy carbon, platinum, and gold. These electrodes are generally encased in a rod of inert insulator with a disk exposed at one end. A regular working electrode has a radius within an order of magnitude of 1 mm. Having a controlled surface area with a well-defined shape is necessary for being able to interpret cyclic voltammetry results.

To run cyclic voltammetry experiments at very high scan rates a regular working electrode is insufficient. High scan rates create peaks with large currents and increased resistances, which result in distortions. Ultramicroelectrodes can be used to minimize the current and resistance.

The counter electrode, also known as the auxiliary or second electrode, can be any material that conducts current easily and will not react with the bulk solution. Reactions occurring at the counter electrode surface are unimportant as long as it continues to conduct current well. To maintain the observed current the counter electrode will often oxidize or reduce the solvent or bulk electrolyte.

Solvents

CV can be conducted using a variety of solutions. Using typical electrodes, solvents dissolve not only the analyte, often at mM levels, but also electrolyte, generally at much higher concentrations. For aqueous solutions, these requirements are trivial, but for nonaqueous solutions, the choices of suitable solvents are fewer.

Electrolyte

The electrolyte ensures good electrical conductivity and minimizes iR drop such that the recorded potentials correspond to actual potentials. For aqueous solutions, many electrolytes are available, but typical ones are alkali metal salts of perchlorate and nitrate. In nonaqueous solvents, the range of electrolytes is more limited, and a popular choice is tetrabutylammonium hexafluorophosphate ("TBAF").

Related Potentiometric Techniques

Potentiodynamic techniques also exist that add low-amplitude AC perturbations to a potential ramp and measure variable response in a single frequency (AC voltammetry) or in many frequencies simultaneously (potentiodynamic electrochemical impedance spectroscopy). The response in alternating current is two-dimensional, characterized by both amplitude and phase. These data can be analyzed to determine information about different chemical processes (charge transfer, diffusion, double layer charging, etc.). Frequency response analysis enables simultaneous monitoring of the various processes that contribute to the potentiodynamic AC response of an electrochemical system.

Whereas cyclic voltammetry is not a hydrodynamic technique, useful electrochemical methods are. In such cases, flow is achieved at the electrode surface by stirring the solution, pumping the solution, or rotating the electrode as is the case with rotating disk electrodes and rotating ring-disk electrodes. Such techniques target steady state conditions and produce waveforms that appear the same when scanned in either the positive or negative directions, thus limiting them to linear sweep voltammetry.

Applications

Cyclic voltammetry (CV) has become an important and widely used electroanalytical technique in many areas of chemistry. It is often used to study a variety of redox processes, to determine the stability of reaction products, the presence of intermediates in redox reactions, reaction. and electron transfer kinetics, and the reversibility of a reaction. CV can also be used to determine the electron stoichiometry of a system, the diffusion coefficient of an analyte, and the formal reduction potential of an analyte, which can be used as an identification tool. In addition, because concentration is proportional to current in a reversible, Nernstian system, the concentration of an unknown solution can be determined by generating a calibration curve of current vs. concentration.

In cellular biology it is used to measure the concentrations in living ones. In organometallic chemistry, it is used to evaluate redox mechanisms.

Measuring Antioxidant Capacity

Cyclical voltammetry can be used to determine the antioxidant capacity in food and even skin. Low molecular weight antioxidants, molecules that prevent other molecules from being oxidized by acting as reducing agents, are important in living cells because they inhibit cell damage or death caused by oxidation reactions that produce radicals. Examples of antioxidants include flavonoids, whose antioxidant activity is greatly increased with more hydroxyl groups. Because traditional methods to determine antioxidant capacity involve tedious steps, techniques to increase the rate of the experiment are continually being researched. One such technique involves cyclic voltammetry because it can measure the antioxidant capacity by quickly measuring the redox behavior over a complex system without the need to measure each component's antioxidant capacity.Furthermore, antioxidants are quickly oxidized at inert electrodes, so the half-wave potential can be utilized to determine antioxidant capacity. It is important to note that whenever cyclic voltammetry is utilized, it is usually compared to spectrophotometry or High Performance Liquid Chromotography (HPLC). Applications of the technique extend to food chemistry, where it is used to determine the antioxidant activity of redwine, chocolate, and hops. Additionally, it even has uses in the world of medicine in that it can determine antioxidants in the skin.

Evaluation of a Technique

The technique being evaluated uses voltammetric sensors combined in an electronic tongue to observe the antioxidant capacity in red wines. These Electronic Tongues (ETs) consist of multiple sensing units like voltammetric sensors, which will have unique responses to certain compounds. This approach is optimal to use since samples of high complexity can be analyzed with high cross-selectivity. Thus, the sensors can be sensitive to pH and antioxidants. As usual, the voltage in the cell was monitored using a working electrode and a reference electrode (Silver/Silver Chloride electrode). Furthermore, a platinum counter electrode allows the current to continue to flow during the experiment. The Carbon Paste Electrodes sensor (CPE) and the Graphite-Epoxy Composite (GEC) electrode are tested in a saline solution before the scanning of the wine so that a reference signal can be obtained. The wines are then ready to be scanned, once with CPE and once with GEC. While cyclic voltammetry was successfully used to generate currents using the wine samples, the signals were complex and needed an additional extraction stage. It was found that the ET method could successfully analyze wine's antioxidant capacity as it agreed with traditional methods like TEAC, Folin-Ciocalteu, and I280 indexes. Additionally, the time was reduced, the sample did not have to be pretreated, and other reagents were unnecessary, all of which diminshed the popularity of traditional methods. Thus, cyclic voltammetry successfully determines the antioxidant capacity and even improves previous results.

Antioxidant Capacity of Chocolate and Hops

The phenolic antioxidants for cocoa powder, dark chocolate, and milk chocolate can also be determined via cyclic voltammetry. In order to achieve this, the anodic peaks are calculated and analyzed with the knowledge that the first and third anodic peaks can be assigned to the first and second oxidation of flavonoids, while the second anodic peak represents phenolic acids. Using the graph produced by cyclic voltammetry, the total phenolic and flavonoid content can be deduced in each of the three samples. It was observed that cocoa powder and dark chocolate had the highest antioxidant capacity since they had high total phenolic and flavonoid content. Milk chocolate had the lowest capacity as it had the lowest phenolic and flavonoid content. While the antioxidant content was given using the cyclic voltammetry anodic peaks, HPLC must then be used to determine the purity of catechins and procyanidin in cocoa powder, dark chocolate, and milk chocolate. Hops, the flowers used in making beer, contain antioxidant properties due to the presence of flavonoids and other polyphenolic compounds.In this cyclic voltammetry experiment, the working electrode voltage was determined using a ferricinium/ferrocene reference electrode. By comparing different hop extract samples, it was observed that the sample containing polyphenols that were oxidized at less positive potentials proved to have better antioxidant capacity.

Anodic Stripping Voltammetry

Anodic stripping voltammetry is a voltammetric method for quantitative determination of specific ionic species. The analyte of interest is electroplated on the working electrode during a deposition step, and oxidized from the electrode during the stripping step. The current is measured during the stripping step. The oxidation of species is registered as a peak in the current signal at the potential at which the species begins to be oxidized. The stripping step can be either linear, staircase, squarewave, or pulse.

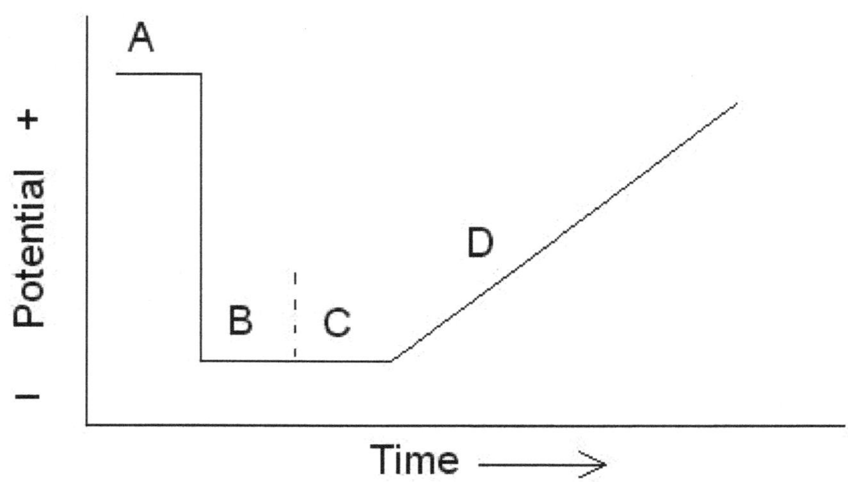

A: Cleaning step, **B:** Electroplating step, **C:** Equilibration step, **D:** Stripping step.

Electrochemical Cell Set-Up

Anodic stripping voltammetry usually incorporates three electrodes, a working electrode, auxiliary electrode (sometimes called the counter electrode), and reference electrode. The solution being analyzed usually has an electrolyte added to it. For most standard tests, the working electrode is a bismuth or mercury film electrode (in a disk or planar strip configuration). The mercury film forms an amalgam with the analyte of interest, which upon oxidation results in a sharp peak, improving resolution between analytes. The mercury film is formed over a glassy carbon electrode. A mercury drop electrode has also been used for much the same reasons. In cases where the analyte of interest has an oxidizing potential above that of mercury, or where a mercury electrode would be otherwise unsuitable, a solid, inert metal such as silver, gold, or platinum may also be used.

Steps

Anodic stripping voltammetry usually incorporates 4 steps if the working electrode is a mercury film or mercury drop electrode and the solution incorporates stirring. The solution is stirred during the first two steps at a repeatable rate. The first step is a cleaning step; in the cleaning step, the potential is held at a more oxidizing potential than the analyte of interest for a period of time in order to fully remove it from the electrode. In the second step, the potential is held at a lower potential, low enough to reduce the analyte and deposit it on the electrode. After the second step, the stirring is stopped, and the electrode is kept at the lower potential. The purpose of this third step is to allow the deposited material to distribute more evenly in the mercury. If a solid inert electrode is used, this step is unnecessary. The last step involves raising the working electrode to a higher potential (anodic), and stripping (oxidizing) the analyte. As the analyte is oxidized, it gives off electrons which are measured as a current.

Stripping analysis is an analytical technique that involves (i) preconcentration of a metal phase onto a solid electrode surface or into Hg (liquid) at negative potentials and (ii) selective oxidation of each metal phase species during an anodic potential sweep.

Stripping analysis has the following properties:

- Very sensitive and reproducible (RSD<5%) method for trace metal ion analysis in aqueous media.
- Concentration limits of detection for many metals are in the low ppb to high ppt range (S/N=3) and this compares favorably with AAS or ICP analysis.
- Field deployable instrumentation that is inexpensive.
- Approximately 12-15 metal ions can be analyzed for by this method.
- The stripping peak currents and peak widths are a function of the size, coverage and distribution of the metal phase on the electrode surface (Hg or alternate)

Sensitivity

Anodic stripping voltammetry can detect **μg**/L concentrations of analyte. This method has an excellent detection limit (typically 10^{-9} - 10^{-10} M)

Applications

Stripping voltammetry (Anodic, Cathodic and Adsorptive) have been employed for analysis of organic molecules as well as metal ions. Carbon paste, glassy carbon paste, glassy carbon etc. electrodes when modified are termed as chemically modified electrodes. Chemically modified electrodes have been employed for the analysis of organic and inorganic compounds.

Cathodic Stripping Voltammetry

Cathodic stripping voltammetry is a voltammetric method for quantitative determination of specific ionic species. It is similar to the trace analysis method anodic stripping voltammetry, except that for the plating step, the potential is held at an oxidizing potential, and the oxidized species are stripped from the electrode by sweeping the potential positively. This technique is used for ionic species that form insoluble salts and will deposit on or near the anodic, working electrode during deposition. The stripping step can be either linear, staircase, squarewave, or pulse.

Adsorptive Stripping Voltammetry

Adsorptive stripping voltammetry is similar to anodic stripping voltammetry and cathodic stripping voltammetry except that the preconcentration step is not controlled by electrolysis. The preconcentration step in adsorptive stripping voltammetry is accomplished by adsorption on the working electrode surface, or by reactions with chemically modified electrodes.

Rotated Electrode Voltammetry

Rotated electrode voltammetry, a subclass of voltammetry, is technique in which the working electrode usually a rotating disk electrode (RDE) or rotating ring-disk electrode (RRDE) is rotated at a very high rate (2000 to 10,000 (?) RPM). This asserts some control over the mass transport process which brings the analyte to the electrode surface. This technique was made popular through the work of the Russian electrochemist, Benjamin Levich. This technique is useful for studying the kinetics and electrochemical reaction mechanism for a half reaction. This technique is part of a larger class of electroanalytical methods known as hydrodynamic voltammetry.

History

The beginning of voltammetry was facilitated by the discovery of polarography in 1922 by the Nobel Prize–winning chemist Jaroslav Heyrovský. Early voltammetric techniques had many problems, limiting their viability for everyday use in analytical chemistry. In polarography, these problems included the fact that Mercury is oxidized at a potential that is more positive than +0.2, which makes it harder to analyze that results for the analytes in the positive region of the potential. Another problem included the residual current obtained from the charging of the large capacitance of the electrode surface. When Heyrovsky first recorded the first dependence on the current flowing through the dropping mercury electrode on the applied potential in 1922, he took point-by-point measurements and he plotted a current-voltage curve. This was considered to be the first polarogram. In order to facilitate this process, he constructed what is now known as a polarograph with M. Shikata and, with enabled him to record photographically the same curve in a matter of hours. He gave recognition to the importance of potential and its control and also recognized the opportunities of measuring the limiting currents. He was also an important part of the introduction of dropping mercury electrode as a modern-day tool.

In 1942 Archie Hickling built the first three electrodes potentiostat, which was an advancement for the field of electrochemistry. He used this potentiostat to control the voltage of an electrode. In the meantime, the late 1940s, Kenneth Cole invented an electronic circuit which he called a voltage clamp. The voltage clamp was used to analyze the ionic conduction in nerves.

The 1960s and 1970s saw many advances in the theory, instrumentation, and the introduction of computer added and controlled systems. Modern polarographic and voltammetric methods on mercury electrodes came about in three sections.

The first section includes the development of the mercury electrodes. The following electrodes were produced: dropping mercury electrode, mercury steaming electrode, hanging mercury drop electrode, static mercury drop electrode, mercury film electrode, mercury amalgam electrodes, mercury microelectrodes, chemically modified mercury electrodes, controlled growth mercury electrodes, and contractible mercury drop electrodes.

There was also an advancement of the measuring techniques used. These measuring techniques include : classical DC polarography, oscillopolarography, Kaloussek's switcher, AC polarography, tast polarography, normal pulse polarography, differential pulse polarography, square-wave voltammetry, cyclic voltammetry, anodic stripping voltammetry, convolution techniques, and elimination methods. Lastly, there was also an advancement of preconcentration techniques that produced an increased the sensitivity of the mercury electrodes. This came about through the development of anodic stripping voltammetry, cathodic stripping voltammetry and adsorptive stripping voltammetry.

These advancements improved sensitivity and created new analytical methods, which prompted the industry to respond with the production of cheaper potentiostat, electrodes, and cells that could be effectively used in routine analytical work.

Applications

Voltammetric Sensors

A number of voltammetric systems are produced commercially for the determination of specific species that are of interest in industry and research. These devices are sometimes called electrodes but are, in fact, complete voltammetric cells and are better referred to as sensors. These sensors can be employed for the analysis of various organic and inorganic analytes in various matrices.

The Oxygen Electrode

The determination of dissolved oxygen in a variety of aqueous environments, such as sea water, blood, sewage, effluents from chemical plants, and soils is of tremendous importance to industry, biomedical and environmental research, and clinical medicine. One of the most common and convenient methods for making such measurements is with the Clark oxygen sensor, which was patented by L.C. Clark, Jr. in 1956.

Polarography

Polarography is a subclass of voltammetry that uses a dropping mercury electrode as the working electrode.

Amperometry

Amperometry is the term indicating the whole of electrochemical techniques in which a current is measured as a function of an independent variable that is, typically, time or electrode potential. Chronoamperometry is the technique in which the current is measured, at a fixed potential, at different times since the start of polarisation. Chronoamperometry is typically carried out in unstirred solution and at fixed electrode, i.e., under experimental conditions avoiding convection as the mass transfer to the electrode. On the other hand, voltammetry is a subclass of amperometry, in which the current is measured by varying the potential applied to the electrode. According to the waveform that describes the way how the potential is varied as a function of time, the different voltammetric techniques are defined. Confusion arose recently about the correct use of many terms proper of electrochemistry/electroanalysis, often owing to the diffusion of electroanalytical techniques in fields where they constitute an instrument to use, not being the 'core business' of the study. Though electrochemists are pleased about this, they invite to use the terms properly, in order to avoid fatal misunderstandings.

REFERENCES

1. Skoog, Douglas A.; Donald M. West; F. James Holler (1995-08-25). Fundamentals of Analytical Chemistry (7th ed.). Harcourt Brace College Publishers. ISBN 0-03-005938-0.
2. Kissinger, Peter; William R. Heineman (1996-01-23). Laboratory Techniques in Electroanalytical Chemistry, Second Edition, Revised and Expanded (2 ed.). CRC. ISBN 0-8247-9445-1.
3. Bard, Allen J.; Larry R. Faulkner (2000-12-18). Electrochemical Methods: Fundamentals and Applications (2 ed.). Wiley. ISBN 0-471-04372-9.
4. Zoski, Cynthia G. (2007-02-07). Handbook of Electrochemistry. Elsevier Science. ISBN 0-444-51958-0.

5. Grundl, Tim (1994-02-01). "A review of the current understanding of redox capacity in natural, disequilibrium systems". Chemosphere. 28 (3): 613–626. doi:10.1016/0045-6535(94)90303-4.

6. Noyhouzer, T.; Valdinger, I.; Mandler, D. (2013-09-03). "Enhanced Potentiometry by Metallic Nanoparticles". Analytical Chemistry. 85 (17): 8347–8353. doi:10.1021/ac401744w. ISSN 0003-2700.

7. H. F. Weber, Wied. Ann., 7, 536, 1879

8. DeFord, Donald D. (1960). "Electroanalysis and Coulometric Analysis". Analytical Chemistry. 32 (5): 31–37. doi:10.1021/ac60161a604.

9. Kissinger, Peter; William R. Heineman (1996-01-23). Laboratory Techniques in Electroanalytical Chemistry, Second Edition, Revised and Expanded (2 ed.). CRC. ISBN 0-8247-9445-1.

10. Zoski, Cynthia G. (2007-02-07). Handbook of Electrochemistry. Elsevier Science. ISBN 0-444-51958-0.

11. Bard, Allen J.; Larry R. Faulkner (2000-12-18). Electrochemical Methods: Fundamentals and Applications (2 ed.). Wiley. ISBN 0-471-04372-9.

12. Gritzner, G.; J. Kuta (1984). "Recommendations on reporting electrode potentials in nonaqueous solvents". Pure Appl. Chem. 56 (4): 461–466. doi:10.1351/pac198456040461. Retrieved 2009-04-17.

13. Nicholson, R. S.; Irving. Shain (1964-04-01). "Theory of Stationary Electrode Polarography. Single Scan and Cyclic Methods Applied to Reversible, Irreversible, and Kinetic Systems.". Analytical Chemistry. 36 (4): 706–723. doi:10.1021/ac60210a007.

14. Kounaves, Samuel. "Voltammetric Techniques". Tufts University Department of Chemistry

15. Barek, J. (2003). "Eighty Years of Polarography - History and Future". Electroanalysis. 15: 467–472.

16. Hickling, A. (1942). "Studies in electrode polarisation. Part IV.-The automatic control of the potential of a working electrode". Transactions of the Faraday Society. 38: 27–33. doi:10.1039/TF9423800027.

17. Sanghavi, Bankim; Srivastava, Ashwini (2010). "Simultaneous voltammetric determination of acetaminophen, aspirin and caffeine using an in situ surfactant-modified multiwalled carbon nanotube paste electrode". Electrochimica Acta. 55: 8638–8648. doi:10.1016/j.electacta.2010.07.093.

CHAPTER 6

Electron Transfer Mechanism

Chapter Contents

- **Introduction**
- **Classes of Electron Transfer**
- **Theory**

INTRODUCTION

Electron transfer (ET) occurs when an electron relocates from an atom or molecule) to another such chemical entity. ET is a mechanistic description of a redox reaction, wherein the oxidation state of reactant and product changes.

Numerous biological processes involve ET reactions. These processes include oxygen binding, photosynthesis, respiration, and detoxification. Additionally, the process of energy transfer can be formalized as a two-electron exchange (two concurrent ET events in opposite directions) in case of small distances between the transferring molecules. ET reactions commonly involve transition metal complexes, but there are now many examples of ET in organic chemistry.

CLASSES OF ELECTRON TRANSFER

There are several classes of electron transfer, defined by the state of the two redox centers and their connectivity

Inner-Sphere Electron Transfer

Inner sphere or bonded electron transfer is a redox chemical reaction that proceeds via a covalent linkage—a strong electronic interaction—between the oxidant and the reductant reactants. In Inner Sphere (IS) electron transfer (ET), a ligand bridges the two metal redox centers during the electron transfer event. Inner sphere reactions are inhibited by large ligands, which prevent the formation of the crucial bridged intermediate. Thus, IS ET is rare in biological systems, where redox sites are often shielded by bulky proteins. Inner sphere ET is usually used to describe reactions involving transition metal complexes and most of this article is written from this perspective. However, redox centers can consist of organic groups rather than metal centers.

The bridging ligand could be virtually any entity that can convey electrons. Typically, such a ligand has more than one lone electron pair, such that it can serve as an electron donor to both the reductant and the oxidant. Common bridging ligands include the halides and the pseudohalides such as hydroxide and thiocyanate. More complex bridging ligands are also well known including oxalate, malonate, and pyrazine. Prior to ET, the bridged complex must form, and such processes are often highly reversible. Electron transfer occurs through the bridge once it is established. In some cases, the stable bridged structure may exist in the ground state; in other cases, the bridged structure may be a transiently-formed intermediate, or else as a transition state during the reaction.

The alternative to inner sphere electron transfer is outer sphere electron transfer. In any transition metal redox process, the mechanism can be assumed to be outer sphere unless the conditions of the inner sphere are met. Inner sphere electron transfer is generally enthalpically more favorable than outer sphere electron transfer due to a larger degree of interaction between the metal centers involved, however, inner sphere electron transfer is usually entropically less favorable since the two sites involved must become more ordered (come together via a bridge) than in outer sphere electron transfer.

Taube's Experiment

The discoverer of the inner sphere mechanism was Henry Taube, who was awarded the Nobel Prize in Chemistry in 1983 for his pioneering studies. A particularly historic finding is summarized in the abstract of the seminal publication. "When $Co(NH_3)_5Cl^{++}$ is reduced by Cr^{++} in M {meaning 1M} $HClO_4$, 1 Cl− appears attached to Cr for each Cr(III) which is formed or Co(III) reduced. When the reaction is carried on in a medium containing radioactive Cl, the mixing of the Cl− attached to Cr(III) with that in solution is less than 0.5%. This experiment shows that transfer of Cl to the reducing agent from the oxidizing agent is direct..." The paper and the excerpt above can be described with the following equation:

$$[CoCl(NH_3)_5]^{2+} + [Cr(H_2O)_6]^{2+} \rightarrow [Co(NH_3)_5(H_2O)]^{2+} + [CrCl(H_2O)_5]^{2+}$$

The point of interest is that the chloride that was originally bonded to the cobalt, the oxidant, becomes bonded to chromium, which in its +3 oxidation state, forms kinetically inert bonds to its ligands. This observation implies the intermediacy of the bimetallic complex $[Co(NH_3)_5(\mu\text{-}Cl)Cr(H_2O)_5]^{4+}$, wherein "μ-Cl" indicates that the chloride bridges between the Cr and Co atoms, serving as a ligand for both. This chloride serves as a conduit for electron flow from Cr(II) to Co(III), forming Cr(III) and Co(II).

The Creutz-Taube Ion

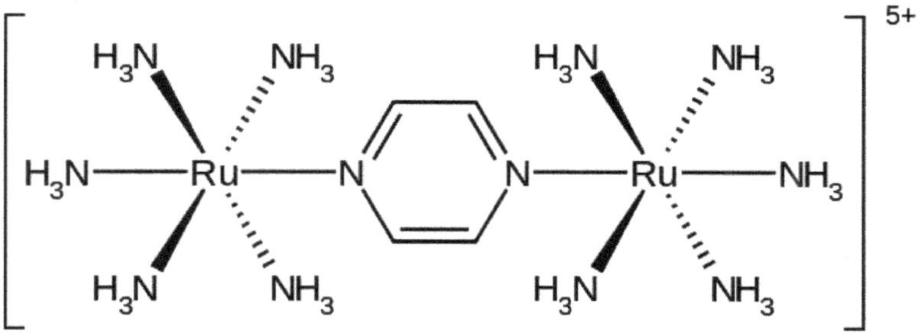

In the preceding example, the occurrence of the chloride bridge is *inferred* from the product analysis, but it was *not* observed. One complex that serves as a model for the bridged intermediate is the "Creutz Taube complex," $[(NH_3)_5RuNC_4H_4NRu(NH_3)_5]^{5+}$. This species is named after Carol Creutz, who prepared the ion during her PhD studies with Henry Taube. The bridging ligand is the heterocycle pyrazine, 1,4-$C_4H_4N_2$. In the Creutz-Taube Ion, the average oxidation state of Ru is 2.5+. Spectroscopic studies, however, show that the two Ru centers are equivalent, which indicates the ease with which the electron hole communicates between the two metals. The significance of the Creutz-Taube ion is its simplicity, which facilitates theoretical analysis, and its high symmetry, which ensures a high degree of delocalization. Many more complex mixed valence species are known both as molecules and polymeric materials.

Mixed Valence Compounds

Mixed valence compounds contain an element which is present in more than one oxidation state. Well-known mixed valence compounds include the Creutz-Taube complex, Prussian blue and Molybdenum blue. Many solids are mixed-valency including indium chalcogenides. Mixed valency is required for organic metals to exhibit electrical conductivity.

As the extinction coefficient decreases, the coupling constant decreases, influencing the angle to increase.

Mixed-valence compounds are subdivided into three groups, according to the Robin-Day Classification:

- Class I, where the valences are "trapped," or localized on a single site, such as Pb_3O_4 and antimony tetroxide. There are distinct sites with different specific valences in the complex that cannot easily interconvert.
- Class II, which are intermediate in character. There is some localization of distinct valences, but there is a low activation energy for their interconversion. Some thermal activation is required to induce electron transfer from one site to another via the bridge. These species exhibit an intense Intervalence charge transfer (IT or IVCT) band, a broad intense absorption in the IR- or visible

part of the spectrum, and also exhibit magnetic exchange coupling at low temperatures. The degree of interaction between the metal sites can be estimated from the absorption profile of the IVCT band and the spacing between the sites. This type of complex is common when metals are in different ligand fields. For example, Prussian blue is an iron(II,III)–cyanide complex in which there is an iron(II) atom surrounded by six carbon atoms of six cyanide ligands bridged to an iron(III) atom by their nitrogen ends. In the Turnbull's blue preparation, an iron(II) solution is mixed with an iron(III) cyanide (c-linked) complex. An electron-transfer reaction occurs via the cyanide ligands to give iron(III) associated with an iron(II)-cyanide complex.

- Class III, wherein mixed valence is not distinguishable by spectroscopic methods as the valence is completely delocalized. The Creutz-Taube Ion is an example of this class of complexes. These species also exhibit an IT band. Each site exhibits an intermediate oxidation state, which can be half-integer in value. This class is possible when the ligand environment is similar or identical for each of the two metal sites in the complex. The bridging ligand needs to be very good at electron transfer, be highly conjugated, and be easily reduced.

Organic mixed valence compounds are also known. Examples are the oxidized form of tetrathiafulvalene and the radical cation of *N,N,N',N'-tetramethyl-p-phenylenediamine*.

Outer-Sphere Electron Transfer

Outer sphere refers to an electron transfer (ET) event that occurs between chemical species that remain separate and intact before, during, and after the ET event. In contrast, for inner sphere electron transfer the participating redox sites undergoing ET become connected by a chemical bridge. Because the ET in outer sphere electron transfer occurs between two non-connected species, the electron is forced to move through space from one redox center to the other.

Marcus Theory

The main theory that describes the rates of outer-sphere electron transfer was developed by Rudolph A. Marcus in the 1950s. A major aspect of Marcus theory is the dependence of the electron transfer rate on the thermodynamic driving force (difference in the redox potentials of the electron-exchanging sites). For most reactions, the rates increase with increased driving force. A second aspect is that the rate of outer-sphere electron-transfer depends inversely on the "reorganizational energy." Reorganization energy describes the changes in bond lengths and angles that are required for the oxidant and reductant to switch their oxidation states. This energy is assessed by measurements of the self-exchange rates (see below). Outer sphere electron transfer is the most common type of electron transfer, especially in biochemistry, where redox centers are separated by several (up to about 11) angstroms by intervening protein. In biochemistry, there are two main types of outer sphere ET: ET between two biological molecules or fixed distance electron transfer, in which the electron transfers within a *single* biomolecule (e.g., intraprotein).

Examples

Self-Exchange

Outer sphere electron transfer can occur between chemical species that are identical except for their oxidation state. This process is termed self-exchange. An example is the degenerate reaction between the tetrahedral ions permanganate and manganate:

$[MnO_4]^- + [Mn^*O_4]^{2-} \rightarrow [MnO_4]^{2-} + [Mn^*O_4]^-$

For octahedral metal complexes, the rate constant for self-exchange reactions correlates with changes the population of the e_g orbitals, the population of which most strongly affects the length of metal-ligand bonds:

- For the $[Co(bipy)_3]^+/[Co(bipy)_3]^{2+}$ pair, self exchange proceeds at 10^9 $M^{-1}s^{-1}$. In this case, the electron configuration changes from Co(I): $(t_{2g})^6(e_g)^2$ to Co(II): $(t_{2g})^5(e_g)^2$.
- For the $[Co(bipy)_3]^{2+}/[Co(bipy)_3]^{3+}$ pair, self exchange proceeds at 18 $M^{-1}s^{-1}$. In this case, the electron configuration changes from Co(II): $(t_{2g})^5(e_g)^2$ to Co(III): $(t_{2g})^6(e_g)^0$.

Iron-Sulfur Proteins

Outer sphere ET is the basis of the biological function of the iron-sulfur proteins. The Fe centers are typically further coordinated by cysteinyl ligands. The $[Fe_4S_4]$ electron-transfer proteins ($[Fe_4S_4]$ ferredoxins) may be further subdivided into low-potential (bacterial-type) and high-potential (HiPIP) ferredoxins. Low- and high-potential ferredoxins are related by the following redox scheme:

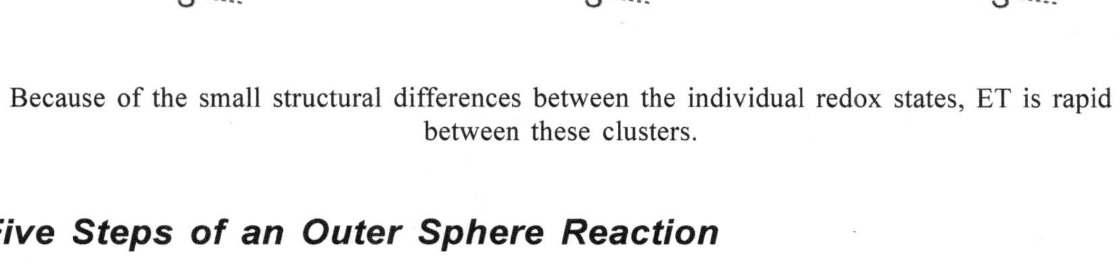

Because of the small structural differences between the individual redox states, ET is rapid between these clusters.

Five Steps of an Outer Sphere Reaction

- 1. reactants diffuse together out of their solvent shells => precursor complex (requires work $=w_r$)
- 2. changing bond lengths, reorganize solvent => activated complex
- 3. Electron transfer

- 4. Relaxation of bond lengths, solvent molecules => successor complex
- 5. Diffusion of products (requires work=w_p)

Heterogeneous Electron Transfer

In heterogeneous electron transfer, an electron moves between a chemical species and a solid-state electrode. Theories addressing heterogeneous electron transfer have applications in electrochemistry and the design of solar cells.

THEORY

The first generally accepted theory of ET was developed by Rudolph A. Marcus to address outer-sphere electron transfer and was based on a transition-state theory approach. The Marcus theory of electron transfer was then extended to include inner-sphere electron transfer by Noel Hush and Marcus. The resultant theory, called Marcus-Hush theory, has guided most discussions of electron transfer ever since. Both theories are, however, semiclassical in nature, although they have been extended to fully quantum mechanical treatments by Joshua Jortner, Alexander M. Kuznetsov, and others proceeding from Fermi's Golden Rule and following earlier work in non-radiative transitions. Furthermore, theories have been put forward to take into account the effects of vibronic coupling on electron transfer; in particular, the PKS theory of electron transfer.

Before 1991, ET in metalloproteins was thought to affect primarily the diffuse, averaged properties of the non-metal atoms forming an insulated barrier between the metals, but Beratan, Betts and Onuchic subsequently showed that the ET rates are governed by the bond structures of the proteins -- that the electrons, in effect, tunnel through the bonds comprising the chain structure of the proteins.

REFERENCES

1. Greenwood, N. N.; & Earnshaw, A. (1997). Chemistry of the Elements (2nd Edn.), Oxford:Butterworth-Heinemann. ISBN 0-7506-3365-4.
2. Holleman, A. F.; Wiberg, E. "Inorganic Chemistry" Academic Press: San Diego, 2001. ISBN 0-12-352651-5.
3. Susan B. Piepho, Elmars R. Krausz, P. N. Schatz; J. Am. Chem. Soc., 1978, 100 (10), pp 2996–3005; Vibronic coupling model for calculation of mixed valence absorption profiles; doi:10.1021/ja00478a011; Publication Date: May 1978
4. Beratan DN, Betts JN, Onuchic JN, Science 31 May 1991: Vol. 252 no. 5010 pp. 1285-1288; Protein electron transfer rates set by the bridging secondary and tertiary structure; doi:10.1126/science.1656523; Publication Date: May 1991
5. IUPAC, Compendium of Chemical Terminology, 2nd ed. (the "Gold Book") (1997). Online corrected version: (2006–) "Inner-sphere electron transfer".
6. Taube, H.; Myers, H.; Rich, R. L. "The Mechanism of Electron Transfer in Solution", Journal of the American Chemical Society, 1953, volume 75, pages 4118-19.doi: 10.1021/ja01112a546

7. Richardson, D. E.; Taube, H., "Mixed-Valence Molecules: Electronic Delocalization and Stabilization", Coordination Chemistry Reviews, 1984, volume 60, pages 107-29.doi:10.1016/0010-8545(84)85063-8

8. Robin, Melvin B.; Day, Peter., "Mixed Valence Chemistry", Advances in Inorganic Chemistry and Radiochemistry, 1967, volume 10, pages 247-422. doi:10.1016/S0065-2792(08)60179-X

9. Optical transitions of symmetrical mixed-valence systems in the Class II-III regime. Bruce S. Brunschwig, Carol Creutz, and Norman Sutin, Chem. Soc. Rev., 2002, 31, pp 168-184.

10. Organic Mixed Valence Jihane Hankache and Oliver S. Wenger Chem. Rev., 2011, 111 (8), pp 5138–5178 doi:10.1021/cr100441k

11. Article: outer-sphere electron transfer, from the IUPAC Gold book]

12. S. J. Lippard, J. M. Berg "Principles of Bioinorganic Chemistry" University Science Books: Mill Valley, CA; 1994. ISBN 0-935702-73-3.

13. R. G. Wilkins Kinetics and Mechanism of Reactions of Transition Metal Complexes, 2nd Edition, VCH, Weinheim, 1991. ISBN 1-56081-125-0

CHAPTER 7

Electrochemical Reaction Mechanism

Chapter Contents

- **Introduction**
- **Formalism**
- **Characterization**
- **Electrocatalysis**
- **History**
- **"Following" Physical Transformations**
- **Electrochemical Potential**
- **Electrochemical Gradient**

INTRODUCTION

In chemistry, an electrochemical reaction mechanism is the step by step sequence of elementary steps, involving at least one outer sphere electron transfer, by which an overall chemical change occurs.

Elementary steps like proton coupled electron transfer and the movement of electrons between an electrode and substrate are special to electrochemical processes. Electrochemical mechanisms are important to all redox chemistry including corrosion, redox active photochemistry including photosynthesis, other biological systems often involving electron transport chains and other forms of homogeneous and heterogeneous electron transfer. Such reactions are most often studied with standard three electrode techniques such as cyclic voltammetry(CV), chronoamperometry, and bulk electrolysis as well as more complex experiments involving rotating disk electrodes and rotating ring-disk electrodes. In the case of photoinduced electron transfer the use of time-resolved spectroscopy is common.

FORMALISM

When describing electrochemical reactions an "E" and "C" formalism is often employed. The E represents an electron transfer; sometimes E_O and E_R are used to represent oxidations and reductions respectively. The C represents a chemical reaction which can be any elementary reaction step and is often called a "following" reaction. In coordination chemistry common C steps which "follow" electron transfer are ligand loss and association. The ligand loss or gain is associated with a geometric change in the complexes coordination sphere.

$[ML_n]^{2+} + e^- \rightarrow [ML_n]^+$ **E**

$[ML_n]^+ \rightarrow [ML_{(n-1)}]^+ + L$ **C**

The reaction above would be called an EC reaction.

CHARACTERIZATION

The production of $[ML_{(n-1)}]^+$ in the reaction above by the "following" chemical reaction produces a species directly at the electrode that could display redox chemistry anywhere in a CV plot or none at all. The change in coordination from $[ML_n]^+$ to $[ML_{(n-1)}]^+$ often prevents the observation of "reversible" behavior during electrochemical experiments like cyclic voltammetry. On the forward scan the expected diffusion wave is observed, in example above the reduction of $[ML_n]^{2+}$ to $[ML_n]^{1+}$. However, on the return scan the corresponding wave is not observed, in the example above this would be the wave corresponding to the oxidation of $[ML_n]^{1+}$ to $[ML_n]^{2+}$. In our example there is no $[ML_n]^{1+}$ to oxidize since it has been converted to $[ML_{(n-1)}]^+$through ligand loss. The return wave can sometimes be observed by increasing the scan rates so the following chemical reaction can be observed before the chemical reaction takes place. This often requires the use of ultramicroelectrodes (UME) capable of very high scan rates of 0.5 to 5.0 V/s. Plots of forward and reverse peak ratios against modified forms of the scan rate often identify the rate of the chemical reaction. It has become a common practice to model such plots with electrochemical simulations. The results of such studies are of disputed practical relevance since simulation requires excellent experimental data, better than that routinely obtained and reported. Furthermore, the parameters of such studies are rarely reported and often include an unreasonably high variable to data ratio (ref?). A better practice is to look for a simple, well documented relationship between observed results and implied phenomena; or to investigate a specific physical phenomenon using an alternative technique such as chronoamperometry or those involving a rotating electrode.

ELECTROCATALYSIS

In chemistry and manufacturing, electrolysis is a technique that uses a direct electric current (DC) to drive an otherwise non-spontaneous chemical reaction. Electrolysis is commercially important as a stage in the separation of elements from naturally occurring sources such as ores using an electrolytic cell. The voltage that is needed for electrolysis to occur is called the decomposition potential.

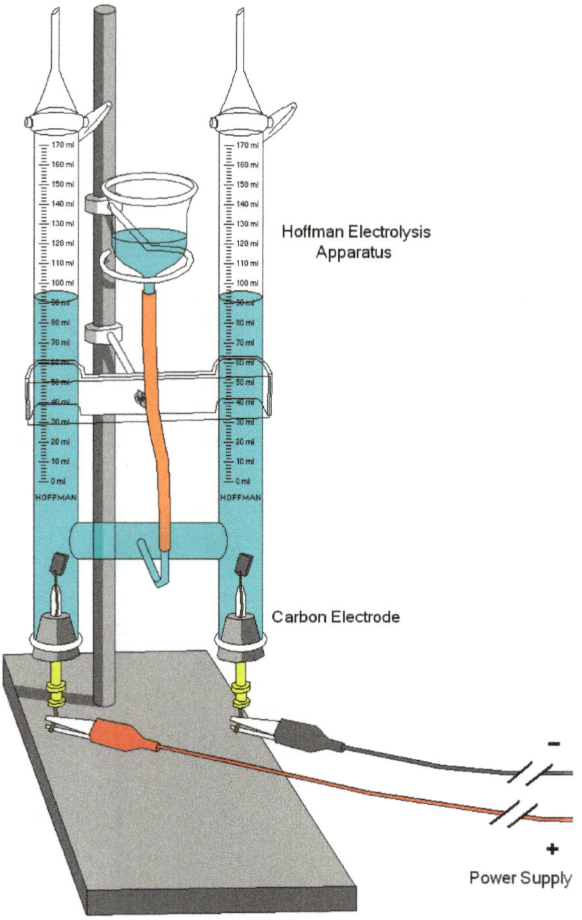

Illustration of an electrolysis apparatus used in a school laboratory.

HISTORY

The word "electrolysis" was introduced by Michael Faraday in the 19th century, on the suggestion of the Rev. William Whewell, using the Greek words lektron amber, which since the 17th century was associated with electric phenomena, and *λύσις* [lýsis] meaning dissolution. Nevertheless, electrolysis, as a tool to study chemical reactions and obtain pure elements, precedes the coinage of the term and formal description by Faraday.

- 1785 – Martinus van Marum›s electrostatic generator was used to reduce tin, zinc, and antimony from their salts using electrolysis.
- 1800 – William Nicholson and Anthony Carlisle (view also Johann Ritter), decomposed water into hydrogen and oxygen.
- 1808 – Potassium (1807), sodium, barium, calcium and magnesium were discovered by Sir Humphry Davy using electrolysis.

- 1821 – Lithium was discovered by William Thomas Brande, who obtained it by electrolysis of lithium oxide.
- 1833 – Michael Faraday develops his two laws of electrolysis, and provides a mathematical explanation of his laws.
- 1875 – Paul Émile Lecoq de Boisbaudran discovered gallium using electrolysis.
- 1886 – Fluorine was discovered by Henri Moissan using electrolysis.
- 1886 – Hall–Héroult process developed for making aluminium
- 1890 – Castner–Kellner process developed for making sodium hydroxide

Overview

Electrolysis is the passing of a direct electric current through an ionic substance that is either molten or dissolved in a suitable solvent, producing chemical reactions at the electrodes and separation of materials.

The main components required to achieve electrolysis are:

- An electrolyte: a substance, frequently an ion-conducting polymer that contains free ions, which carry electric current in the electrolyte. If the ions are not mobile, as in a solid salt then electrolysis cannot occur.
- A direct current (DC) electrical supply: provides the energy necessary to create or discharge the ions in the electrolyte. Electric current is carried by electrons in the external circuit.
- Two electrodes: electrical conductors that provide the physical interface between the electrolyte and the electrical circuit that provides the energy.

Electrodes of metal, graphite and semiconductor material are widely used. Choice of suitable electrode depends on chemical reactivity between the electrode and electrolyte and manufacturing cost.

Process of Electrolysis

The key process of electrolysis is the interchange of atoms and ions by the removal or addition of electrons from the external circuit. The desired products of electrolysis are often in a different physical state from the electrolyte and can be removed by some physical processes. For example, in the electrolysis of brine to produce hydrogen and chlorine, the products are gaseous. These gaseous products bubble from the electrolyte and are collected.

$$2\ NaCl + 2\ H_2O \rightarrow 2\ NaOH + H_2 + Cl_2$$

A liquid containing mobile ions (electrolyte) is produced by:

- Solvation or reaction of an ionic compound with a solvent (such as water) to produce mobile ions
- An ionic compound is fused by heating

An electrical potential is applied across a pair of electrodes immersed in the electrolyte.

Each electrode attracts ions that are of the opposite charge. Positively charged ions (cations) move towards the electron-providing (negative) cathode. Negatively charged ions (anions) move towards the electron-extracting (positive) anode.

In this process electrons are either absorbed or released. Neutral atoms gain or lose electrons and become charged ions that then pass into the electrolyte. The formation of uncharged atoms from ions is called discharging. When an ion gains or loses enough electrons to become uncharged (neutral) atoms, the newly formed atoms separate from the electrolyte. Positive metal ions like Cu^{++} deposit onto the cathode in a layer. The terms for this are electroplating, electrowinning, and electrorefining. When an ion gains or loses electrons without becoming neutral, its electronic charge is altered in the process. In chemistry, the loss of electrons is called Oxidation, while electron gain is called reduction.

Oxidation and Reduction at the Electrodes

Oxidation of ions or neutral molecules occurs at the anode. For example, it is possible to oxidize ferrous ions to ferric ions at the anode:

$Fe_2+(aq) \rightarrow Fe_3+(aq) + e^-$

Reduction of ions or neutral molecules occurs at the cathode.

It is possible to reduce ferricyanide ions to ferrocyanide ions at the cathode:

$Fe(CN)_3\text{-}6 + e^- \rightarrow Fe(CN)_4\text{-}6$

Neutral molecules can also react at either of the electrodes. For example: p-Benzoquinone can be reduced to hydroquinone at the cathode:

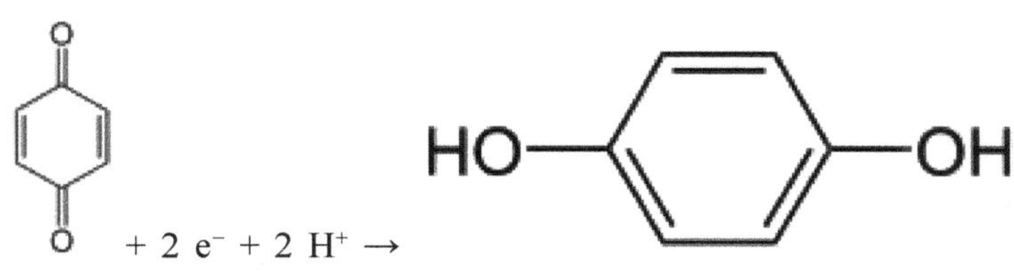

In the last example, H^+ ions (hydrogen ions) also take part in the reaction, and are provided by an acid in the solution, or by the solvent itself (water, methanol etc.). Electrolysis reactions involving H^+ ions are fairly common in acidic solutions. In aqueous alkaline solutions, reactions involving OH^- (hydroxide ions) are common.

Sometimes the solvents themselves (usually water) are oxidized or reduced at the electrodes. It is even possible to have electrolysis involving gases. Such as when using a Gas diffusion electrode.

Energy Changes During Electrolysis

The amount of electrical energy that must be added equals the change in Gibbs free energy of the reaction plus the losses in the system. The losses can (in theory) be arbitrarily close to zero, so the maximum thermodynamic efficiency equals the enthalpy change divided by the free energy change of the reaction. In most cases, the electric input is larger than the enthalpy change of the reaction, so some energy is released in the form of heat. In some cases, for instance, in the electrolysis of steam into hydrogen and oxygen at high temperature, the opposite is true and heat energy is absorbed. This heat is absorbed from the surroundings, and the heating value of the produced hydrogen is higher than the electric input.

Related Techniques

The following techniques are related to electrolysis:

- Electrochemical cells, including the hydrogen fuel cell, use differences in Standard electrode potential to generate an electrical potential that provides useful power. Though related via the interaction of ions and electrodes, electrolysis and the operation of electrochemical cells are quite distinct. However, a chemical cell should not be seen as performing electrolysis in reverse.

Faraday's Laws of Electrolysis

First Law of Electrolysis

In 1832, Michael Faraday reported that the quantity of elements separated by passing an electric current through a molten or dissolved salt is proportional to the quantity of electric charge passed through the circuit. This became the basis of the first law of electrolysis:

$$m = k \cdot q$$

or

$$m = eQ$$

where; e is known as electrochemical equivalent of the metal deposited or of the gas liberated at the electrode.

Second Law of Electrolysis

Faraday discovered that when the same amount of current is passed through different electrolytes/elements connected in series, the mass of substance liberated/deposited at the electrodes is directly proportional to their equivalent weight.

Industrial uses

- Electrometallurgy is the process of reduction of metals from metallic compounds to obtain the pure form of metal using electrolysis. aluminium, lithium, sodium, potassium, magnesium, calcium, and in some cases copper, are produced in this way.
- Production of chlorine and sodium hydroxide
- Production of sodium chlorate and potassium chlorate

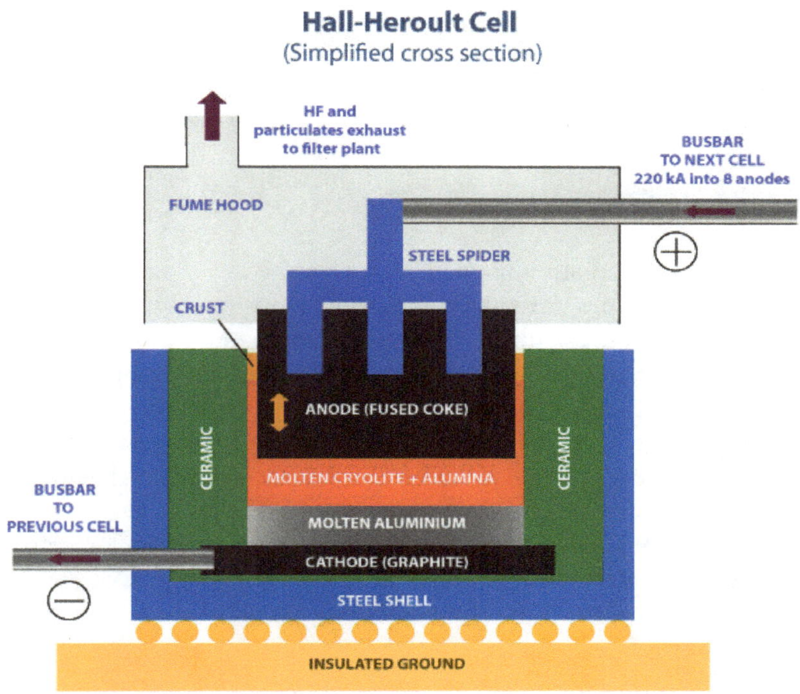

Hall-Heroult process for producing aluminium.

- Production of perfluorinated organic compounds such as trifluoroacetic acid by the process of electrofluorination
- Production of electrolytic copper as a cathode, from refined copper of lower purity as an anode.

Electrolysis has many other uses:

- Production of oxygen for spacecraft and nuclear submarines.
- Production of hydrogen for fuel, using a cheap source of electrical energy.

Electrolysis is also used in the cleaning and preservation of old artifacts. Because the process separates the non-metallic particles from the metallic ones, it is very useful for cleaning a wide variety of metallic objects, from old coins to even larger objects including rusted cast iron cylinder blocks and heads when rebuilding automobile engines. Rust removal from small iron or steel objects by electrolysis can be done in a home workshop using simple materials such as a plastic bucket, tap water, lengths of rebar, washing soda, baling wire, and a battery charger.

Manufacturing Processes

In manufacturing, electrolysis can be used for:

- Electroplating, where a thin film of metal is deposited over a substrate material. Electroplating is used in many industries for either functional or decorative purposes, as in vehicle bodies and nickel coins.

- Electrochemical Machining (ECM), where an electrolytic cathode is used as a shaped tool for removing material by anodic oxidation from a workpiece. ECM is often used as technique for deburring or for etching metal surfaces like tools or knives with a permanent mark or logo.

Competing Half-Reactions in Solution Electrolysis

Using a cell containing inert platinum electrodes, electrolysis of aqueous solutions of some salts leads to reduction of the cations (e.g., metal deposition with, e.g., zinc salts) and oxidation of the anions (e.g. evolution of bromine with bromides). However, with salts of some metals (e.g. sodium) hydrogen is evolved at the cathode, and for salts containing some anions (e.g. sulfate SO_4^{2-}) oxygen is evolved at the anode. In both cases this is due to water being reduced to form hydrogen or oxidized to form oxygen. In principle the voltage required to electrolyze a salt solution can be derived from the standard electrode potential for the reactions at the anode and cathode. The standard electrode potential is directly related to the Gibbs free energy, ΔG, for the reactions at each electrode and refers to an electrode with no current flowing. An extract from the table of standard electrode potentials is shown below.

Half-reaction	***E*° (V)**	**Ref.**
$Na^+ + e^- \rightleftharpoons Na(s)$	−2.71	
$Zn^{2+} + 2e^- \rightleftharpoons Zn(s)$	−0.7618	
$2H^+ + 2e^- \rightleftharpoons H_2(g)$	**≡ 0**	
$Br_2(aq) + 2e^- \rightleftharpoons 2Br^-$	+1.0873	
$O_2(g) + 4H^+ + 4e^- \rightleftharpoons 2H_2O$	+1.23	
$Cl_2(g) + 2e^- \rightleftharpoons 2Cl^-$	+1.36	
$S_2O^{2-}8 + 2e^- \rightleftharpoons 2SO2-4$	+2.07	

In terms of electrolysis, this table should be interpreted as follows:

- Oxidized species (often a cation) with a more negative cell potential are more difficult to reduce than oxidized species with a more positive cell potential. For example, it is more difficult to reduce a sodium ion to a sodium metal than it is to reduce a zinc ion to a zinc metal.
- Reduced species (often an anion) with a more positive cell potential are more difficult to oxidize than reduced species with a more negative cell potential. For example, it is more difficult to oxidize sulfate anions than it is to oxidize bromide anions.

Using the Nernst equation the electrode potential can be calculated for a specific concentration of ions, temperature and the number of electrons involved. For pure water (pH 7):

- the electrode potential for the reduction producing hydrogen is −0.41 V
- the electrode potential for the oxidation producing oxygen is +0.82 V.

Comparable figures calculated in a similar way, for 1M zinc bromide, $ZnBr_2$, are −0.76 V for the reduction to Zn metal and +1.10 V for the oxidation producing bromine. The conclusion from these figures is that hydrogen should be produced at the cathode and oxygen at the anode from the electrolysis of water—which is at variance with the experimental observation that zinc metal is deposited and bromine is produced. The explanation is that these calculated potentials only indicate the thermodynamically preferred reaction. In practice many other factors have to be taken into account such as the kinetics of some of the reaction steps involved. These factors together mean that a higher potential is required for the reduction and oxidation of water than predicted, and these are termed overpotentials. Experimentally it is known that overpotentials depend on the design of the cell and the nature of the electrodes.

For the electrolysis of a neutral (pH 7) sodium chloride solution, the reduction of sodium ion is thermodynamically very difficult and water is reduced evolving hydrogen leaving hydroxide ions in solution. At the anode the oxidation of chlorine is observed rather than the oxidation of water since the overpotential for the oxidation of chloride to chlorine is lower than the overpotential for the oxidation of water to oxygen. The hydroxide ions and dissolved chlorine gas react further to form hypochlorous acid. The aqueous solutions resulting from this process is called electrolyzed water and is used as a disinfectant and cleaning agent.

Research Trends

Electrolysis of Carbon Dioxide

The electrochemical reduction or electrocatalytic conversion of CO_2 can produce value-added chemicals such methane, ethylene, ethane, etc. The electrolysis of carbon dioxide gives formate or carbon monoxide, but sometimes more elaborate organic compounds such as ethylene. This technology is under research as a carbon-neutral route to organic compounds.

Electrolysis of Water

Electrolysis of water produces hydrogen.

$2\ H_2O(l) \rightarrow 2\ H_2(g) + O_2(g);\ E_0 = +1.229\ V$

The energy efficiency of water electrolysis varies widely. The efficiency of an electrolyser is a measure of the enthalpy contained in the hydrogen (to undergo combustion with oxygen, or some other later reaction), compared with the input electrical energy. Heat/enthalpy values for hydrogen are well published in science and engineering texts, as 144 MJ/kg. Note that fuel cells (not electrolysers) cannot utilise this full amount of heat/enthalpy, which has led to some confusion when calculating efficiency values for both types of technology. In the reaction, some energy is lost as heat. Some reports quote efficiencies between 50% and 70% for alkaline electrolysers; however, much higher practical efficiencies are available with the use of PEM (Polymer Electrolyte Membrane electrolysis) and catalytic technology, such as 95% efficiency.

NREL estimated that 1 kg of hydrogen (roughly equivalent to 3 kg, or 4 L, of petroleum in energy terms) could be produced by wind powered electrolysis for between \$5.55 in the near term and \$2.27 in the long term.

About 4% of hydrogen gas produced worldwide is generated by electrolysis, and normally used onsite. Hydrogen is used for the creation of ammonia for fertilizer via the Haber process, and converting heavy petroleum sources to lighter fractions via hydrocracking.

Carbon/Hydrocarbon Assisted Water Electrolysis

Recently, to reduce the energy input, the utilization of carbon (coal), alcohols (hydrocarbon solution), and organic solution (glycerol, formic acid, ethylene glycol, etc.) with co-electrolysis of water has been proposed as a viable option. The carbon/hydrocarbon assisted electrolysis for hydrogen generation would performs this operation in a single electrochemical reactor. This system energy balance can be required only around 40% electric input with 60% coming from the chemical energy of carbon or hydrocarbon.

Electrocrystallization

A specialized application of electrolysis involves the growth of conductive crystals on one of the electrodes from oxidized or reduced species that are generated in situ. The technique has been used to obtain single crystals of low-dimensional electrical conductors, such as charge-transfer salts.

"FOLLOWING" PHYSICAL TRANSFORMATIONS

During potential variant experiments common to go through a redox couple in which the major species is transformed from a species that is soluble in the solution to one that is insoluble. This results in nucleation process in which a new species plates out on the working electrode. If a species has been deposited on the electrode during a potential sweep then on the return sweep a stripping wave is usually observed.

$[ML_n]^+_{(solvated)} + e^- \rightarrow [ML_n]^0_{(solid)}$ nucleation

$[ML_n]^0_{(solid)} \rightarrow e^- + [ML_n]^+_{(solvated)}$ stripping

While the nucleation wave may be pronounced or difficult the detect the stripping wave is usually very distinct. Often these phenomenon can be avoided by reducing the concentration of the complex in solution. Neither these physical state changes involve a chemical reaction mechanism but they are worth mentioning here since the resulting data is at times confused with some chemical reaction mechanisms.

ELECTROCHEMICAL POTENTIAL

In electrochemistry, the electrochemical potential, μ, sometimes abbreviated to ECP, is a thermodynamic measure of chemical potential that does not omit the energy contribution of electrostatics. Electrochemical potential is expressed in the unit of J/mol.

Each chemical species (for example, water molecules, sodium ions, electrons, etc.) has an electrochemical potential (a quantity with units of energy) at any given point in space, which represents how easy or difficult it is to add more of that species to that location. If possible, a species will move from areas with higher electrochemical potential to areas with lower electrochemical potential; in equilibrium, the electrochemical potential will be constant everywhere for each species (it may have a different value for different species). For example, if a glass of water has sodium ions (Na^+) dissolved uniformly in it, and an electric field is applied across the water, then the sodium ions will tend to get pulled by the electric field towards one side. We say the ions have electric potential energy, and are moving to lower their potential energy. Likewise, if a glass of water has a lot of dissolved sugar on one side and none on the other side, each sugar molecule will randomly diffuse around the water, until there is equal concentration of sugar everywhere. We say that the sugar molecules have a "chemical potential," which is higher in the high-concentration areas, and the molecules move to lower their chemical potential. These two examples show that an electrical potential and a chemical potential can both give the same result: A redistribution of the chemical species. Therefore, it makes sense to combine them into a single "potential", the electrochemical potential, which can directly give the net redistribution taking both into account.

It is (in principle) easy to measure whether or not two regions (for example, two glasses of water) have the same electrochemical potential for a certain chemical species (for example, a solute molecule): Allow the species to freely move back and forth between the two regions (for example, connect them with a semi-permeable membrane that lets only that species through). If the chemical potential is the same in the two regions, the species will occasionally move back and forth between the two regions, but on average there is just as much movement in one direction as the other, and there is zero net migration (this is called "diffusive equilibrium"). If the chemical potentials of the two regions are different, more molecules will move to the lower chemical potential than the other direction.

Moreover, when there is not diffusive equilibrium, i.e., when there is a tendency for molecules to diffuse from one region to another, then there is a certain free energy released by each net-diffusing molecule. This energy, which can sometimes be harnessed (a simple example is a concentration cell), and the free-energy per mole is exactly equal to the electrochemical potential difference between the two regions.

Conflicting Terminologies

It is common in electrochemistry and solid-state physics to discuss both the chemical potential and the electrochemical potential of the electrons. However, in the two fields, the definitions of these two terms are sometimes swapped. In electrochemistry, the electrochemical potential of electrons (or any other species) is the total potential, including both the (internal, nonelectrical) chemical potential and the electric potential, and is by definition constant across a device in equilibrium, whereas the chemical potential of electrons is equal to the electrochemical potential minus the local electric potential energy per electron. In solid-state physics, the definitions are normally compatible with this, but occasionally the definitions are swapped.

This article uses the electrochemistry definitions.

Definition and Usage

In generic terms, electrochemical potential is the mechanical work done in bringing 1 mole of an ion from a standard state to a specified concentration and electrical potential. According to the IUPAC definition, it is the partial molar Gibbs energy of the substance at the specified electric potential, where the substance is in a specified phase. Electrochemical potential can be expressed as

$$\bar{\mu}_i = \mu_i + z_i F \Phi,$$

where:

- μ_i is the electrochemical potential of species i, in J/mol
- μ_i is the chemical potential of the species i, in J/mol
- z_i is the valency (charge) of the ion i, a dimensionless integer
- F is the Faraday constant, in C/mol
- Φ is the local electrostatic potential, in V.

In the special case of an uncharged atom, $z_i = 0$ and so $\mu_i = \mu_i$.

Electrochemical potential is important in biological processes that involve molecular diffusion across membranes, in electroanalytical chemistry, and industrial applications such as batteries and fuel cells. It represents one of the many interchangeable forms of potential energy through which energy may be conserved.

In cell membranes, the electrochemical potential is the sum of the chemical potential and the membrane potential.

Incorrect Usage

The term electrochemical potential is sometimes used to mean an electrode potential (either of a corroding electrode, an electrode with a non-zero net reaction or current, or an electrode at equilibrium). In some contexts, the electrode potential of corroding metals is called "electrochemical corrosion potential", which is often abbreviated as ECP, and the word "corrosion" is sometimes omitted. This usage can lead to confusion. The two quantities have different meanings and different dimensions: the dimension of electrochemical potential is energy per mole while that of electrode potential is voltage (energy per charge).

Fermi Level

The Fermi level is the total chemical potential for electrons (or electrochemical potential for electrons) and is usually denoted by μ or E_F. The Fermi level of a body is a thermodynamic quantity, and its significance is the thermodynamic work required to add one electron to the body (not counting the work required to remove the electron from wherever it came from). A precise understanding of the Fermi level—how it relates to electronic band structure in determining electronic properties, how it relates to the voltage and flow of charge in an electronic circuit—is essential to an understanding of solid-state physics.

In a band structure picture, the Fermi level can be considered to be a hypothetical energy level of an electron, such that at thermodynamic equilibrium this energy level would have a 50% probability of being occupied at any given time. The position of the Fermi level with the relation to the band energy levels is a crucial factor in determining electrical properties. The Fermi level does not necessarily correspond to an actual energy level (in an insulator the Fermi level lies in the band gap), nor does it require the existence of a band structure. Nonetheless, the Fermi level is a precisely defined thermodynamic quantity, and differences in Fermi level can be measured simply with a voltmeter.

The Fermi Level and Voltage

this is not exactly true. As a counterexample, multi-material devices such as p–n junctions contain internal electrostatic potential differences at equilibrium, yet without any accompanying net current; if a voltmeter is attached to the junction, one simply measures zero volts. Clearly, the electrostatic potential is not the only factor influencing the flow of charge in a material—Pauli repulsion, carrier concentration gradients, electromagnetic induction, and thermal effects also play an important role.

A voltmeter measures differences in Fermi level divided by electron charge.

Sometimes it is said that electric currents are driven by differences in electrostatic potential (Galvani potential), but

In fact, the quantity called voltage as measured in an electronic circuit has a simple relationship to the chemical potential for electrons (Fermi level). When the leads of a voltmeter are attached to two points in a circuit, the displayed voltage is a measure of the total work transferred when a unit charge is allowed to move from one point to the other. If a simple wire is connected between two points of differing voltage (forming a short circuit), current will flow from positive to negative voltage, converting the available work into heat.

The Fermi level of a body expresses the work required to add an electron to it, or equally the work obtained by removing an electron. Therefore, $V_A - V_B$, the observed difference in voltage between two points, A and B, in an electronic circuit is exactly related to the corresponding chemical potential difference, $\mu_A - \mu_B$, in Fermi level by the formula

$$V_A - V_B = \frac{\mu_A - \mu_B}{-e}$$

where −e is the electron charge.

From the above discussion it can be seen that electrons will move from a body of high μ (low voltage) to low μ (high voltage) if a simple path is provided. This flow of electrons will cause the lower μ to increase (due to charging or other repulsion effects) and likewise cause the higher μ to decrease. Eventually, μ will settle down to the same value in both bodies. This leads to an important fact regarding the equilibrium (off) state of an electronic circuit:

An electronic circuit in thermodynamic equilibrium will have a constant Fermi level throughout its connected parts.

This also means that the voltage (measured with a voltmeter) between any two points will be zero, at equilibrium. Note that thermodynamic equilibrium here requires that the circuit be internally connected and not contain any batteries or other power sources, nor any variations in temperature.

The Fermi Level and Band Structure

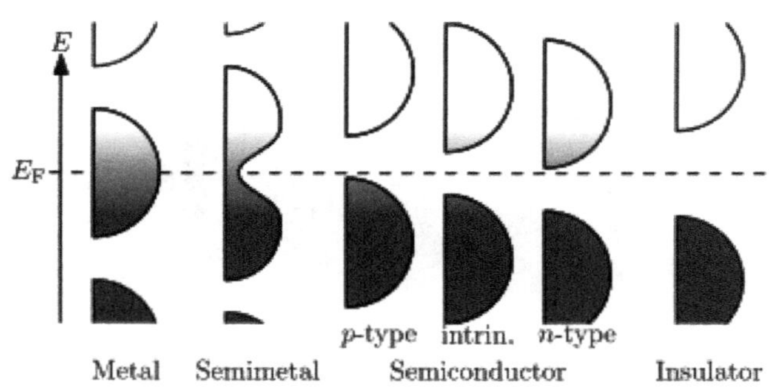

Filling of the electronic states in various types of materials at equilibrium. Here, height is energy while width is the density of available states for a certain energy in the material listed. The shade follows the Fermi–Dirac distribution (**black** = all states filled, **white** = no state filled). In metals and semimetals the **Fermi level** E_F lies inside at least one band. In insulators and semiconductors the Fermi level is inside a band gap; however, in semiconductors the bands are near enough to the Fermi level to be thermally populated with electrons or holes.

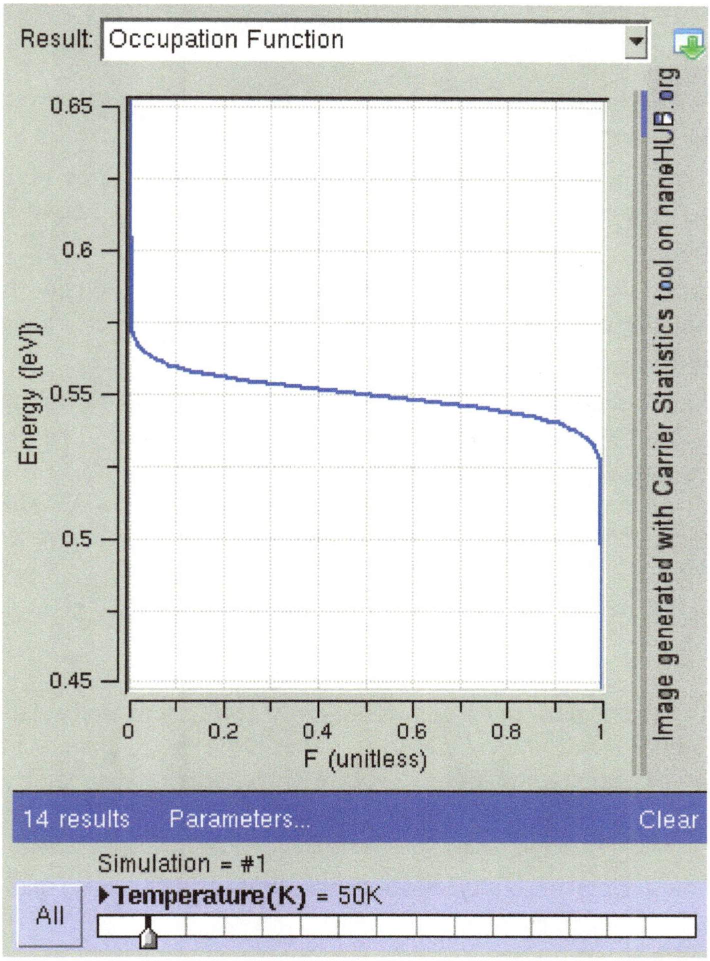

Fermi function F(ε) vs. energy ε, with μ = 0.55 eV and for various temperatures in the range 50K ≤ T ≤ 375K.

In the band theory of solids, electrons are considered to occupy a series of bands composed of single-particle energy eigenstates each labelled by ϵ. Although this single particle picture is an approximation, it greatly simplifies the understanding of electronic behavior and it generally provides correct results when applied correctly.

The Fermi–Dirac distribution, f(ε), gives the probability that (at thermodynamic equilibrium) a state having energy ϵ is occupied by an electron:

$$f(\epsilon) = \frac{1}{e^{(\epsilon-\mu)/kT} + 1}$$

Here, T is the absolute temperature and k is Boltzmann's constant. If there is a state at the Fermi level (ϵ = μ), then this state will have a 50% chance of being occupied. The distribution is plotted in the left figure. The closer f is to 1, the higher chance this state is occupied. The closer f is to 0, the higher chance this state is empty.

The location of μ within a material›s band structure is important in determining the electrical behavior of the material.

- In an insulator, μ lies within a large band gap, far away from any states that are able to carry current.
- In a metal, semimetal or degenerate semiconductor, μ lies within a delocalized band. A large number of states nearby μ are thermally active and readily carry current.
- In an intrinsic or lightly doped semiconductor, μ is close enough to a band edge that there are a dilute number of thermally excited carriers residing near that band edge.

In semiconductors and semimetals the position of μ relative to the band structure can usually be controlled to a significant degree by doping or gating. These controls do not change μ which is fixed by the electrodes, but rather they cause the entire band structure to shift up and down (sometimes also changing the band structure›s shape). For further information about the Fermi levels of semiconductors, see (for example) Sze.

Local Conduction Band Referencing, Internal Chemical Potential and the Parameter ζ

If the symbol $\mathcal{E}$ is used to denote an electron energy level measured relative to the energy of the edge of its enclosing band, ϵ_C, then in general we have $\mathcal{E} = \epsilon - \epsilon_C$. We can define a parameter ζ that references the Fermi level with respect to the band edge:

$$\zeta = \mu - \epsilon_C.$$

It follows that the Fermi–Dirac distribution function can be written as

$$f(\mathcal{E}) = \frac{1}{e^{(\mathcal{E}-\zeta)/kT}+1}.$$

The band theory of metals was initially developed by Sommerfeld, from 1927 onwards, who paid great attention to the underlying thermodynamics and statistical mechanics. Confusingly, in some contexts the band-referenced quantity ζ may be called the Fermi level, chemical potential, or electrochemical potential, leading to ambiguity with the globally-referenced Fermi level. In this article, the terms conduction-band referenced Fermi level or internal chemical potential are used to refer to ζ.

ζ is directly related to the number of active charge carriers as well as their typical kinetic energy, and hence it is directly involved in determining the local properties of the material (such as electrical conductivity). For this reason it is common to focus on the value of ζ when concentrating on the properties of electrons in a single, homogeneous conductive material. By analogy to the energy states of a free electron, the $\mathcal{E}$ of a state is the kinetic energy of that state and ϵ_C is its potential energy. With this in mind, the parameter, ζ, could also be labelled the Fermi kinetic energy.

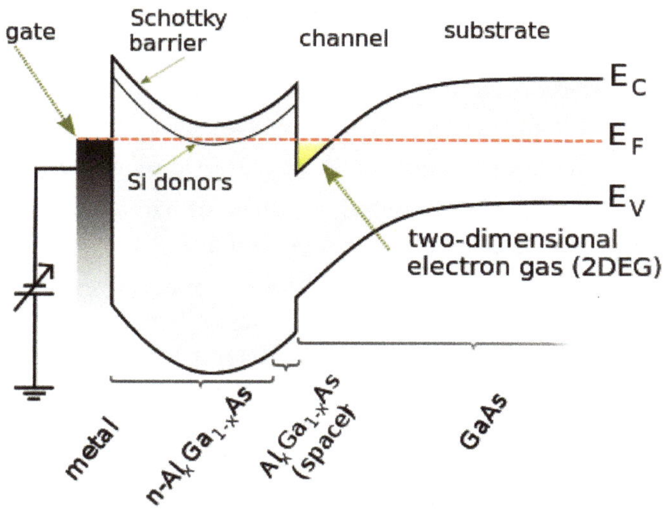

Example of variations in conduction band edge E_C in a band diagram of GaAs/AlGaAs heterojunction-based high-electron-mobility transistor.

Unlike μ, the parameter, ζ, is not a constant at equilibrium, but rather varies from location to location in a material due to variations in ϵ_C, which is determined by factors such as material quality and impurities/dopants. Near the surface of a semiconductor or semimetal, ζ can be strongly controlled by externally applied electric fields, as is done in a field effect transistor. In a multi-band material, ζ may even take on multiple values in a single location. For example, in a piece of aluminum metal there are two conduction bands crossing the Fermi level (even more bands in other materials); each band has a different edge energy, ϵ_C, and a different ζ.

The value of ζ at zero temperature is widely known as the Fermi energy, sometimes written ζ_0. Confusingly (again), the name Fermi energy sometimes is used to refer to ζ at non-zero temperature.

The Fermi Level and Temperature Out of Equilibrium

The Fermi level, μ, and temperature, T, are well defined constants for a solid-state device in thermodynamic equilibrium situation, such as when it is sitting on the shelf doing nothing. When the device is brought out of equilibrium and put into use, then strictly speaking the Fermi level and temperature are no longer well defined. Fortunately, it is often possible to define a quasi-Fermi level and quasi-temperature for a given location, that accurately describe the occupation of states in terms of a thermal distribution. The device is said to be in quasi-equilibrium when and where such a description is possible.

The quasi-equilibrium approach allows one to build a simple picture of some non-equilibrium effects as the electrical conductivity of a piece of metal (as resulting from a gradient of μ) or its thermal conductivity (as resulting from a gradient in T). The quasi-μ and quasi-T can vary (or not exist at all) in any non-equilibrium situation, such as:

- If the system contains a chemical imbalance (as in a battery).

- If the system is exposed to changing electromagnetic fields. (as in capacitors, inductors, and transformers).
- Under illumination from a light-source with a different temperature, such as the sun (as in solar cells),
- When the temperature is not constant within the device (as in thermocouples),
- When the device has been altered, but has not had enough time to re-equilibrate (as in piezoelectric or pyroelectric substances).

In some situations, such as immediately after a material experiences a high-energy laser pulse, the electron distribution cannot be described by any thermal distribution. One cannot define the quasi-Fermi level or quasi-temperature in this case; the electrons are simply said to be non-thermalized. In less dramatic situations, such as in a solar cell under constant illumination, a quasi-equilibrium description may be possible but requiring the assignment of distinct values of μ and T to different bands (conduction band vs. valence band). Even then, the values of μ and T may jump discontinuously across a material interface (e.g., p–n junction) when a current is being driven, and be ill-defined at the interface itself.

Technicalities

Terminology Problems

The term, Fermi level, is mainly used in discussing the solid state physics of electrons in semiconductors, and a precise usage of this term is necessary to describe band diagrams in devices comprising different materials with different levels of doping. In these contexts, however, one may also see Fermi level used imprecisely to refer to the band-referenced Fermi level, $\mu - \epsilon_C$, called ζ above. It is common to see scientists and engineers refer to controlling, pinning, or tuning the Fermi level inside a conductor, when they are in fact describing changes in ϵ_C due to doping or the field effect. In fact, thermodynamic equilibrium guarantees that the Fermi level in a conductor is always fixed to be exactly equal to the Fermi level of the electrodes; only the band structure (not the Fermi level) can be changed by doping or the field effect. A similar ambiguity exists between the terms, chemical potential and electrochemical potential.

It is also important to note that Fermi level is not necessarily the same thing as Fermi energy. In the wider context of quantum mechanics, the term Fermi energy usually refers to the maximum kinetic energy of a fermion in an idealized non-interacting, disorder free, zero temperature Fermi gas. This concept is very theoretical (there is no such thing as a non-interacting Fermi gas, and zero temperature is impossible to achieve). However, it finds some use in approximately describing white dwarfs, neutron stars, atomic nuclei, and electrons in a metal. On the other hand, in the fields of semiconductor physics and engineering, Fermi energy often is used to refer to the Fermi level described in this article.

Fermi Level Referencing and the Location of Zero Fermi Level

Much like the choice of origin in a coordinate system, the zero point of energy can be defined arbitrarily. Observable phenomena only depend on energy differences. When comparing distinct bodies, however, it is important that they all be consistent in their choice of the location of zero energy, or else nonsensical results will be obtained. It can therefore be helpful to explicitly name a common point to ensure that different components are in agreement. On the other hand, if a reference point is inherently ambiguous (such as “the vacuum”, see below) it will instead cause more problems.

A practical and well-justified choice of common point is a bulky, physical conductor, such as the electrical ground or earth. Such a conductor can be considered to be in a good thermodynamic equilibrium and so its μ is well defined. It provides a reservoir of charge, so that large numbers of electrons may be added or removed without incurring charging effects. It also has the advantage of being accessible, so that the Fermi level of any other object can be measured simply with a voltmeter.

Why it is not Advisable To Use “The Energy In Vacuum” As A Reference Zero

In principle, one might consider using the state of a stationary electron in the vacuum as a reference point for energies. This approach is not advisable unless one is careful to define exactly where the vacuum is. The problem is that not all points in the vacuum are equivalent.

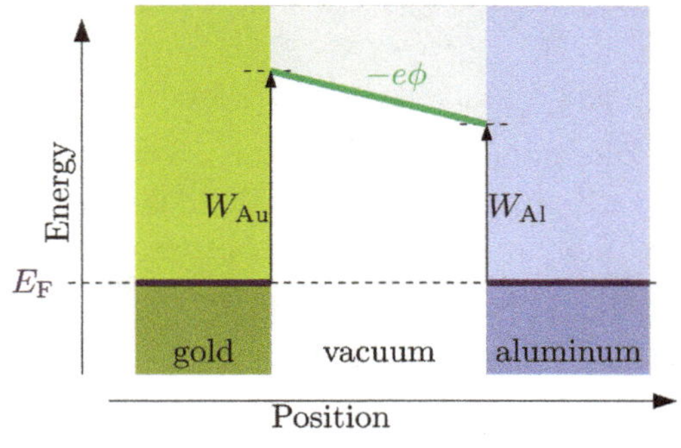

When the two metals depicted here are in thermodynamic equilibrium as shown (equal Fermi levels E_F), the vacuum electrostatic potential ϕ is not flat due to a difference in work function.

At thermodynamic equilibrium, it is typical for electrical potential differences of order 1 V to exist in the vacuum (Volta potentials). The source of this vacuum potential variation is the variation in work function between the different conducting materials exposed to vacuum. Just outside a conductor, the electrostatic potential depends sensitively on the material, as well as which surface is selected (its crystal orientation, contamination, and other details).

The parameter that gives the best approximation to universality is the Earth-referenced Fermi level suggested above. This also has the advantage that it can be measured with a voltmeter.

Discrete Charging Effects in Small Systems

In cases where the "charging effects" due to a single electron are non-negligible, the above definitions should be clarified. For example, consider a capacitor made of two identical parallel-plates. If the capacitor is uncharged, the Fermi level is the same on both sides, so one might think that it should take no energy to move an electron from one plate to the other. But when the electron has been moved, the capacitor has become (slightly) charged, so this does take a slight amount of energy. In a normal capacitor, this is negligible, but in a nano-scale capacitor it can be more important.

In this case one must be precise about the thermodynamic definition of the chemical potential as well as the state of the device: is it electrically isolated, or is it connected to an electrode?

- When the body is able to exchange electrons and energy with an electrode (reservoir), it is described by the grand canonical ensemble. The value of chemical potential μ can be said to be fixed by the electrode, and the number of electrons N on the body may fluctuate. In this case, the chemical potential of a body is the infinitesimal amount of work needed to increase the average number of electrons by an infinitesimal amount (even though the number of electrons at any time is an integer, the average number varies continuously.):

$$\mu(\langle N\rangle, T) = \left(\frac{\partial F}{\partial \langle N\rangle}\right)_T,$$

where F(N, T) is the free energy function of the grand canonical ensemble.

- If the number of electrons in the body is fixed (but the body is still thermally connected to a heat bath), then it is in the canonical ensemble. We can define a "chemical potential" in this case literally as the work required to add one electron to a body that already has exactly N electrons,

$$\mu'(N,T) = F(N+1,T) - F(N,T),$$

where F(N, T) is the free energy function of the canonical ensemble, or alternatively as the work obtained by removing an electron from that body,

$$\mu''(N,T) = F(N,T) - F(N-1,T) = \mu'(N-1,T).$$

These chemical potentials are not equivalent, μ ≠ μ' ≠ μ", except in the thermodynamic limit. The distinction is important in small systems such as those showing Coulomb blockade. The parameter, μ, (i.e., in the case where the number of electrons is allowed to fluctuate) remains exactly related to the voltmeter voltage, even in small systems. To be precise, then, the Fermi level is defined not by a deterministic charging event by one electron charge, but rather a statistical charging event by an infinitesimal fraction of an electron.

ELECTROCHEMICAL GRADIENT

An electrochemical gradient is a gradient of electrochemical potential, usually for an ion that can move across a membrane. The gradient consists of two parts, the chemical gradient, or difference in solute concentration across a membrane, and the electrical gradient, or difference in charge across a membrane. When there are unequal concentrations of an ion across a permeable membrane, the ion will move across the membrane from the area of higher concentration to the area of lower concentration through simple diffusion. Ions also carry an electric charge that forms an electric potential across a membrane. If there is an unequal distribution of charges across the membrane, then the difference in electric potential generates a force that drives ion diffusion until the charges are balanced on both sides of the membrane.

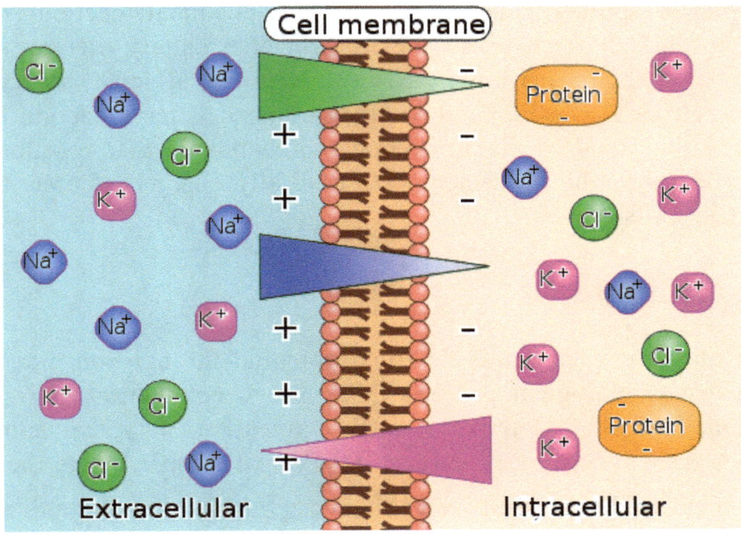

Diagram of ion concentrations and charge across a semi-permeable cellular membrane.

Overview

Electrochemical potential is important in electroanalytical chemistry and industrial applications such as batteries and fuel cells. It represents one of the many interchangeable forms of potential energy through which energy may be conserved.

In biological processes, the direction an ion moves by diffusion or active transport across a membrane is determined by the electrochemical gradient. In mitochondria and chloroplasts, proton gradients are used to generate a chemiosmotic potential that is also known as a proton motive force. This potential energy is used for the synthesis of ATP by oxidative phosphorylation or photophosphorylation, respectively.

An electrochemical gradient has two components. First, the electrical component is caused by a charge difference across the lipid membrane. Second, a chemical component is caused by a differential concentration of ions across the membrane. The combination of these two factors determines the thermodynamically favourable direction for an ion's movement across a membrane.

An electrochemical gradient is analogous to the water pressure across a hydroelectric dam. Membrane transport proteins such as the sodium-potassium pump within the membrane are equivalent to turbines that convert the water's potential energy to other forms of physical or chemical energy, and the ions that pass through the membrane are equivalent to water that ends up at the bottom of the dam. Also, energy can be used to pump water up into the lake above the dam. In similar manner, chemical energy in cells can be used to create electrochemical gradients.

Chemistry

The term is typically applied in contexts wherein a chemical reaction is to take place, such as one involving the transfer of an electron at a battery electrode. In a battery, an electrochemical potential arising from the movement of ions balances the reaction energy of the electrodes. The maximum voltage that a battery reaction can produce is sometimes called the standard electrochemical potential of that reaction. In instances pertaining specifically to the movement of electrically charged solutes, the potential is often expressed in units of volts.

Biological Context

The generation of a transmembrane electrical potential through ion movement across a cell membrane drives biological processes like nerve conduction, muscle contraction, hormonesecretion, and sensory processes. By convention, a typical animal cell has a transmembrane electrical potential between -50mV and -70mV inside the cell.

Electrochemical gradients also play a role in establishing proton gradients in oxidative phosphorylation in mitochondria.The final step of cellular respiration is the electron transport chain. Four complexes embedded in the inner membrane of the mitochondrion make up the electron transport chain. However, only complexes I, III, and IV pump protons from the matrix to the intermembrane space (IMS). In total, there are ten protons translocated from the matrix to the IMS which generates an electrochemical potential of more than 200mV. This drives the flux of protons back into the matrix through ATP synthase which produces ATP by adding an inorganic phosphate to ADP. Thus, generation of a proton electrochemical gradient is crucial for energy production in mitochondria. The total equation for the electron transport chain is:

$$NADH + 11H^+(matrix) + 1/2\,O_2 \longrightarrow NAD^+ + 10H^+(IMS) + H_2O$$

Similar to the electron transport chain, the light-dependent reactions of photosynthesis pump protons into the thylakoid lumen of chloroplasts to drive the synthesis of ATP by ATP synthase. The proton gradient can be generated through either noncyclic or cyclic photophosphorylation. Of the proteins that participate in noncyclic photophosphorylation, photosystem II (PSII), plastiquinone, and cytochrome b_6f complex directly contribute to generating the proton gradient. For each four photons absorbed by PSII, eight protons are pumped into the lumen. The total equation for photophosphorylation is shown:

$$2H_2O + 6H^+(stroma) + 2NADP^+ \longrightarrow O_2 + 8H^+(lumen) + 2NADPH$$

Several other transporters and ion channels play a role in generating a proton electrochemical gradient. One is TPK_3, a potassium channel that is activated by Ca^{2+} and conducts K^+ from the thylakoid lumen to the stroma which helps establish the pH gradient. On the other hand, the electro-neutral K^+ efflux antiporter (KEA_3) transports K^+ into the thylakoid lumen and H^+into the stroma which helps establish the electric field.

Ion Gradients

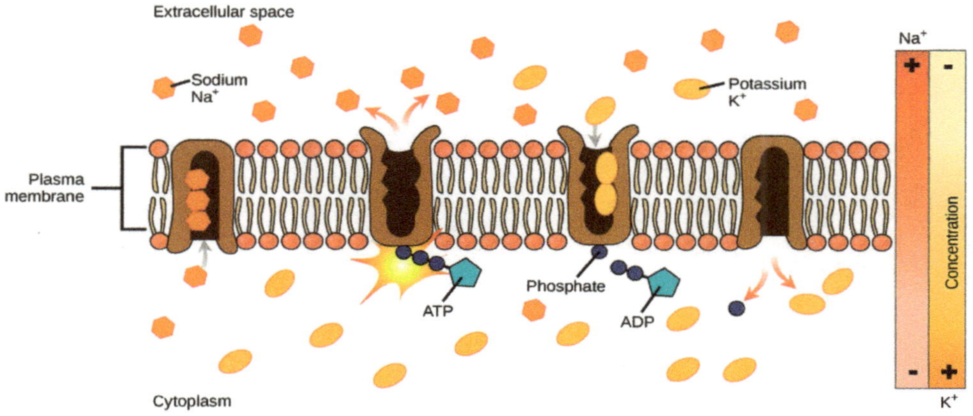

Diagram of the Na^+-K^+-ATPase.

Since the ions are charged, they cannot pass through the membrane via simple diffusion. Two different mechanisms can transport the ions across the membrane: active or passive transport. An example of active transport of ions is the Na^+-K^+-ATPase (NKA). NKA catalyzes the hydrolysis of ATP into ADP and an inorganic phosphate and for every molecule of ATP hydrolized, three Na^+ are transported outside and two K^+ are transported inside the cell. This makes the inside of the cell more negative than the outside and more specifically generates a V_m of about -60mV. An example of passive transport is ion fluxes through Na^+, K^+, Ca^{2+}, and Cl^- channels. These ions tend to move down their concentration gradient. For example, since there is a high concentration of Na^+ outside the cell, Na^+ will flow through the Na^+ channel into the cell. Since the electric potential inside the cell is negative, the influx of a positive ion depolarizes the membrane which brings the transmembrane electric potential closer to zero. However, Na^+ will continue moving down its concentration gradient as long as the effect of the chemical gradient is greater than the effect of the electrical gradient. Once the effect of both gradients are equal (for Na^+ this at a V_m of about +70mV), the influx of Na^+ stops because the driving force (ΔG) is zero. The equation for the driving force is:

$$\Delta G = RTln(C_{in}/C_{out}) + ZFV_m$$

In this equation, R represents the gas constant, T represents absolute temperature, Z is the ionic charge, and F represents the Faraday constant.

Proton Gradients

Proton gradients in particular are important in many different types of cells as a form of energy storage. The gradient is usually used to drive ATP synthase, flagellar rotation, or transport of metabolites. This section will focus on three processes that help establish proton gradients in their respective cells: bacteriorhodopsin and noncyclic photophosphorylation and oxidative phosphorylation.

Bacteriorhodopsin

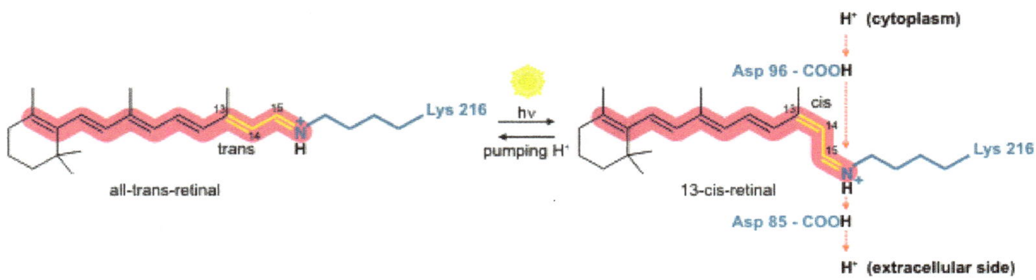

Diagram of the conformational shift in retinal that initiates proton pumping in bacteriorhodopsin.

The way bacteriorhodopsin generates a proton gradient in Archaea is through a proton pump. The proton pump relies on proton carriers to drive protons from the side of the membrane with a low H^+ concentration to the side of the membrane with a high H^+ concentration. In bacteriorhodopsin, the proton pump is activated by absorption of photons of 568 nm wavelength which leads to isomerization of the Schiff base (SB) in retinal forming the K state. This moves SB away from Asp85 and Asp212, causing H^+ transfer from the SB to Asp85 forming the M1 state. The protein then shifts to the M2 state by separating Glu204 from Glu194 which releases a proton from Glu204 into the external medium. The SB is reprotonated by Asp96 which forms the N state. It is important that the second proton comes from Asp96 since its deprotonated state is unstable and rapidly reprotonated with a proton from the cytosol. The protonation of Asp85 and Asp96 causing re-isomerization of the SB forming the O state. Finally, bacteriorhodopsin returns to its resting state when Asp85 releases its proton to Glu204.

Photophosphorylation

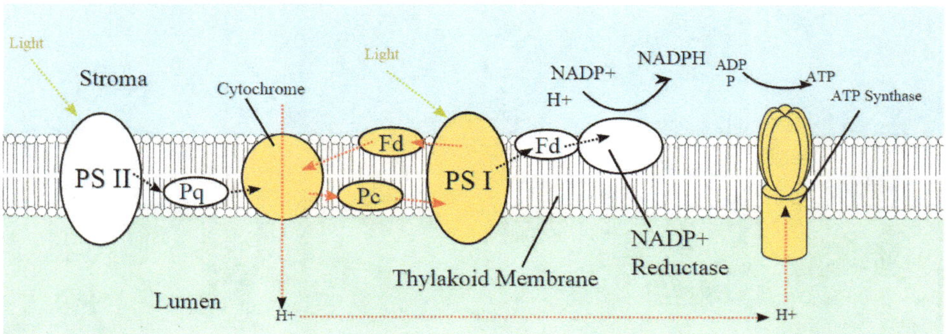

Simplified diagram of photophosphorylation.

PSII also relies on light to drive the formation of proton gradients in chloroplasts, however PSII utilizes vectorial redox chemistry to achieve this goal. Rather than physically transporting protons through the protein, reactions requiring the binding of protons will occur on the extracellular side while reactions requiring the release of protons will occur on the intracellular side. Absorption of photons of 680 nm wavelength is used to excite two electrons in P_{680} to a higher energy level. These higher energy electrons are transferred to protein-bound plastoquinone (PQ_A) and then to unbound plastoquinone (PQ_B). This reduces plastoquinone (PQ) to plastoquinol (PQH_2) which is released from PSII after gaining two protons from the stroma. The electrons in P_{680} are replenished by oxidizing water through the oxygen-evolving complex (OEC). This results in release of O_2 and H^+ into the lumen. The total reaction is shown:

$$4\,photons(680nm) + 2H_2O + 2PQ + 4H^+(stroma) \longrightarrow O_2 + 2PQH_2 + 4H^+(lumen)$$

After being released from PSII, PQH_2 travels to the cytochrome b_6f complex which then transfers two electrons from PQH_2 to plastocyanin in two separate reactions. The process that occurs is similar to the Q-cycle in Complex III of the electron transport chain. In the first reaction, PQH_2 binds to the complex on the lumen side and one electron is transferred to the iron-sulfur center which then transfers it to cytochrome f which then transfers it to plastocyanin. The second electron is transferred to heme b_L which then transfers it to heme b_H which then transfers it to PQ. In the second reaction, a second PQH_2 gets oxidized, adding an electron to another plastocyanin and PQ.Both reactions together transfer four protons into the lumen.

Oxidative Phosphorylation

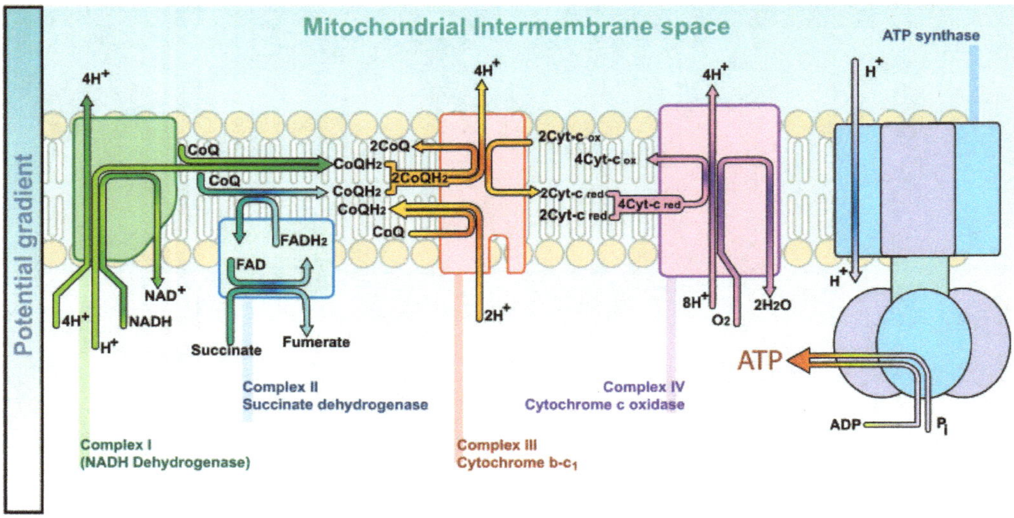

Detailed diagram of the electron transport chain in mitochondria.

In the electron transport chain, Complex I (CI) catalyzes the reduction of ubiquinone (UQ) to ubiquinol (UQH_2) by the transfer of two electrons from reduced nicotinamide adenine dinucleotide (NADH) which translocates four protons from the mitochondrial matrix to the IMS:

$$NADH + H^+ + UQ + 4H^+(matrix) \longrightarrow NAD^+ + UQH_2 + 4H^+(IMS)$$

Complex III (CIII) catalyzes the Q-cycle. The first step involving the transfer of two electrons from the UQH_2 reduced by CI to two molecules of oxidized cytochrome c at the Q_o site. In the second step, two more electrons reduce UQ to UQH_2 at the Q_i site. The total reaction is shown:

$$2\ cytochrome\ c\ (oxidized) + UQH_2 + 2H^+\ (matrix) \longrightarrow 2\ cytochrome\ c\ (reduced) + UQ + 4H^+\ (IMS)$$

Complex IV (CIV) catalyzes the transfer of two electrons from the cytochrome c reduced by CIII to one half of a full oxygen. Utilizing one full oxygen in oxidative phosphorylation requires the transfer of four electrons. The oxygen will then consume four protons from the matrix to form water while another four protons are pumped into the IMS. The total reaction is shown:

$$2\ cytochrome\ c\ (reduced) + 4H^+\ (matrix) + 1/2\ O_2 \longrightarrow 2\ cytochrome\ c\ (oxidized) + 2H^+\ (IMS) + H_2O$$

DIFFUSION

Molecular diffusion, often simply called diffusion, is the thermal motion of all (liquid or gas) particles at temperatures above absolute zero. The rate of this movement is a function of temperature, viscosity of the fluid and the size (mass) of the particles. Diffusion explains the net flux of molecules from a region of higher concentration to one of lower concentration. Once the concentrations are equal the molecules continue to move, but since there is no concentration gradient the process of molecular diffusion has ceased and is instead governed by the process of self-diffusion, originating from the random motion of the molecules. The result of diffusion is a gradual mixing of material such that the distribution of molecules is uniform. Since the molecules are still in motion, but an equilibrium has been established, the end result of molecular diffusion is called a "dynamic equilibrium". In a phase with uniform temperature, absent external net forces acting on the particles, the diffusion process will eventually result in complete mixing.

Consider two systems; S_1 and S_2 at the same temperature and capable of exchanging particles. If there is a change in the potential energy of a system; for example $\mu_1 > \mu_2$ (μ is Chemical potential) an energy flow will occur from S_1 to S_2, because nature always prefers low energy and maximum entropy.

Molecular diffusion is typically described mathematically using Fick's laws of diffusion.

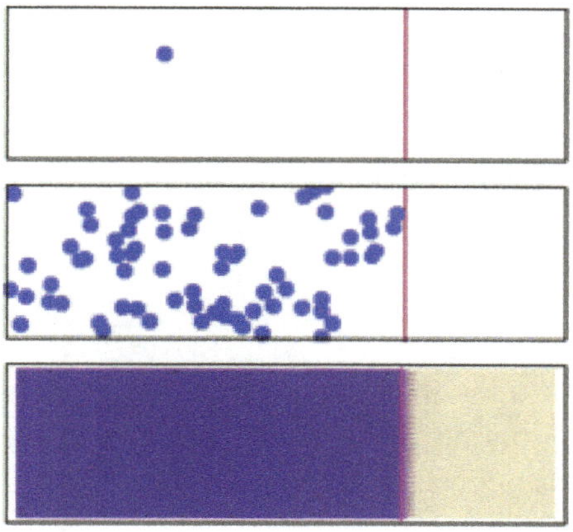

Diffusion from a microscopic and macroscopic point of view. Initially, there are solute molecules on the left side of a barrier (purple line) and none on the right. The barrier is removed, and the solute diffuses to fill the whole container. Top: A single molecule moves around randomly. Middle: With more molecules, there is a clear trend where the solute fills the container more and more uniformly. Bottom: With an enormous number of solute molecules, all randomness is gone: The solute appears to move smoothly and systematically from high-concentration areas to low-concentration areas, following Fick's laws.

Applications

Diffusion is of fundamental importance in many disciplines of physics, chemistry, and biology. Some example applications of diffusion:

- Sintering to produce solid materials (powder metallurgy, production of ceramics)
- Chemical reactor design
- Catalyst design in chemical industry

- Steel can be diffused (e.g., with carbon or nitrogen) to modify its properties
- Doping during production of semiconductors.

Significance

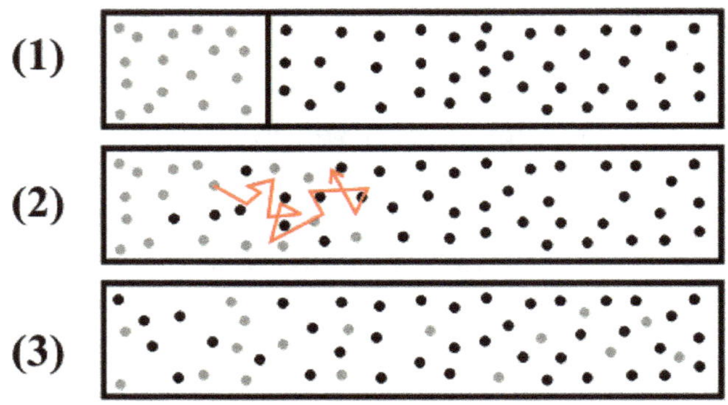

Schematic representation of mixing of two substances by diffusion.

Diffusion is part of the transport phenomena. Of mass transport mechanisms, molecular diffusion is known as a slower one.

Biology

In cell biology, diffusion is a main form of transport for necessary materials such as amino acids within cells. Diffusion of solvents, such as water, through a semipermeable membrane is classified as osmosis.

Metabolism and respiration rely in part upon diffusion in addition to bulk or active processes. For example, in the alveoli of mammalianlungs, due to differences in partial pressures across the alveolar-capillary membrane, oxygen diffuses into the blood and carbon dioxide diffuses out. Lungs contain a large surface area to facilitate this gas exchange process.

Tracer, Self- and Chemical Diffusion

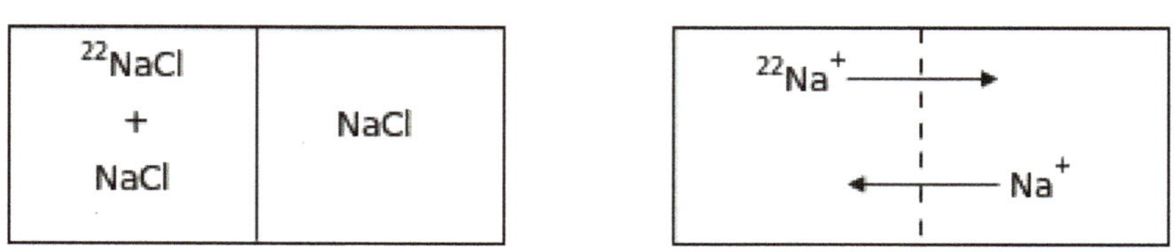

Self diffusion, exemplified with an isotopic tracer of radioactive isotope ^{22}Na.

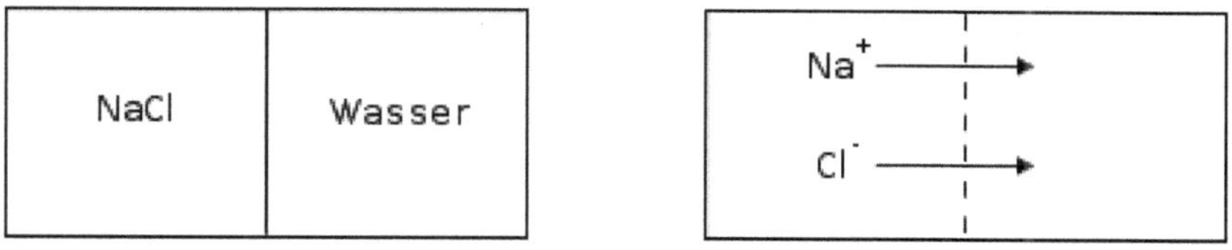

Example of chemical (classical, Fick's, or Fickian) diffusion of sodium chloride in water.

Fundamentally, two types of diffusion are distinguished:

- Tracer diffusion and Self-diffusion, which is a spontaneous mixing of molecules taking place in the absence of concentration (or chemical potential) gradient. This type of diffusion can be followed using isotopic tracers, hence the name. The tracer diffusion is usually assumed to be identical to self-diffusion (assuming no significant isotopic effect). This diffusion can take place under equilibrium. An excellent method for the measurement of self-diffusion coefficients is pulsed field gradient (PFG) NMR, where no isotopic tracers are needed. In a so-called NMR spin echo experiment this technique uses the nuclear spin precession phase, allowing to distinguish chemically and physically completely identical species e.g. in the liquid phase, as for example water molecules within liquid water. The self-diffusion coefficient of water has been experimentally determined with high accuracy and thus serves often as a reference value for measurements on other liquids. The self-diffusion coefficient of neat water is: $2.299 \cdot 10^{-9}$ $m^2 \cdot s^{-1}$ at 25 °C and $1.261 \cdot 10^{-9}$ $m^2 \cdot s^{-1}$ at 4 °C.
- Chemical diffusion occurs in a presence of concentration (or chemical potential) gradient and it results in net transport of mass. This is the process described by the diffusion equation. This diffusion is always a non-equilibrium process, increases the system entropy, and brings the system closer to equilibrium.

The diffusion coefficients for these two types of diffusion are generally different because the diffusion coefficient for chemical diffusion is binary and it includes the effects due to the correlation of the movement of the different diffusing species.

Non-Equilibrium System

Because chemical diffusion is a net transport process, the system in which it takes place is not an equilibrium system (i.e. it is not at rest yet). Many results in classical thermodynamics are not easily applied to non-equilibrium systems. However, there sometimes occur so-called quasi-steady states, where the diffusion process does not change in time, where classical results may locally apply. As the name suggests, this process is a not a true equilibrium since the system is still evolving.

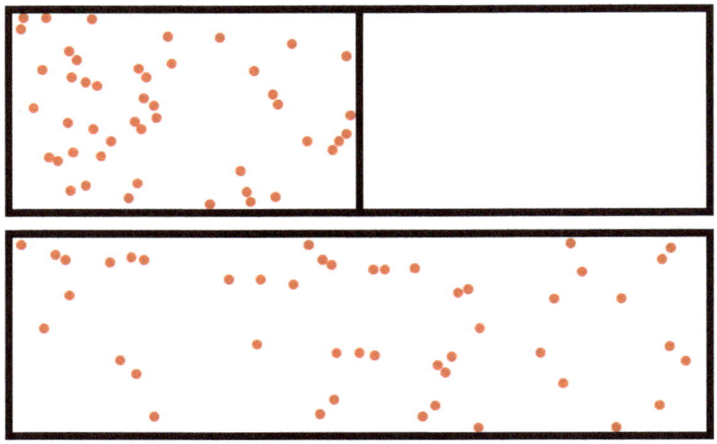

Illustration of low entropy (top) and high entropy (bottom).

Non-equilibrium fluid systems can be successfully modeled with Landau-Lifshitz fluctuating hydrodynamics. In this theoretical framework, diffusion is due to fluctuations whose dimensions range from the molecular scale to the macroscopic scale.

Chemical diffusion increases the entropy of a system, i.e. diffusion is a spontaneous and irreversible process. Particles can spread out by diffusion, but will not spontaneously re-order themselves (absent changes to the system, assuming no creation of new chemical bonds, and absent external forces acting on the particle).

Concentration Dependent "Collective" Diffusion

Contrary to brownian motion, which is the diffusion of a single particle, interactions between particles may have to be considered, unless the particles form an ideal mix with their solvent (ideal mix conditions correspond to the case where the interactions between the solvent and particles are identical to the interactions between particles and the interactions between solvent molecules; in this case, the particles do not interact when inside the solvent).

In case of an ideal mix, the particle diffusion equation holds true and the diffusion coefficient D the speed of diffusion in the particle diffusion equation is independent of particle concentration. In other cases, resulting interactions between particles within the solvent will account for the following effects:

- the diffusion coefficient D in the particle diffusion equation becomes dependent of concentration. For an attractive interaction between particles, the diffusion coefficient tends to decrease as concentration increases. For a repulsive interaction between particles, the diffusion coefficient tends to increase as concentration increases.
- In the case of an attractive interaction between particles, particles exhibit a tendency to coalesce and form clusters if their concentration lies above a

certain threshold. This is equivalent to a precipitation chemical reaction (and if the considered diffusing particles are chemical molecules in solution, then it is a precipitation).

Molecular Diffusion of Gases

Transport of material in stagnant fluid or across streamlines of a fluid in a laminar flow occurs by molecular diffusion. Two adjacent compartments separated by a partition, containing pure gases A or B may be envisaged. Random movement of all molecules occurs so that after a period molecules are found remote from their original positions. If the partition is removed, some molecules of A move towards the region occupied by B, their number depends on the number of molecules at the point considered. Concurrently, molecules of B diffuse toward regimens formerly occupied by pure A. Finally, complete mixing occurs. Before this point in time, a gradual variation in the concentration of A occurs along an axis, designated x, which joins the original compartments. This variation, expressed mathematically as $-dC_A/dx$, where C_A is the concentration of A. The negative sign arises because the concentration of A decreases as the distance x increases. Similarly, the variation in the concentration of gas B is $-dC_B/dx$. The rate of diffusion of A, N_A, depend on concentration gradient and the average velocity with which the molecules of A moves in the x direction. This relationship is expressed by Fick's Law

$$N_A = -D_{AB}\frac{dC_A}{dx}$$ (only applicable for no bulk motion)

where D is the Diffusivity of A through B, proportional to the average (squared?) molecular velocity and, therefore dependent on the temperature and pressure of gases. The rate of Diffusion N_A,is usually expressed as the number of moles diffusing across unit area in unit time. As with the basic equation of heat transfer, this indicates that the rate of force is directly proportional to the driving force, which is the concentration gradient.

This basic equation applies to a number of situations. Restricting discussion exclusively to steady state conditions, in which neither dC_A/dx or dC_B/dx change with time, equimolecular counterdiffusion is considered first.

Equimolecular Counterdiffusion

If no bulk flow occurs in an element of length dx, the rates of diffusion of two gases A and B must be equal and opposite, that is $N_A = -N_B$.

The partial pressure of A changes by dP_A over the distance dx. Similarly, the partial pressure of B changes dP_B. As there is no difference in total pressure across the element (no bulk flow), we have

$$\frac{dP_A}{dx} = -\frac{dP_B}{dx}.$$

For an ideal gas the partial pressure is related to the molar concentration by the relation

$$P_A V = n_A RT$$

where n_A is the number of moles of gas A in a volume V. As the molar concentration C_A is equal to n_A/V therefore

$$P_A = C_A RT$$

Consequently, for gas A,

$$N_A = -D_{AB}\frac{1}{RT}\frac{dP_A}{dx}$$

where D_{AB} is the diffusivity of A in B. Similarly,

$$N_B = -D_{BA}\frac{1}{RT}\frac{dP_B}{dx} = D_{AB}\frac{1}{RT}\frac{dP_A}{dx}$$

Considering that $dP_A/dx=-dP_B/dx$, it therefore proves that $D_{AB}=D_{BA}=D$. If the partial pressure of A at x_1 is P_{A1} and x_2 is P_{A2}, integration of above equation,

$$N_A = -\frac{D}{RT}\frac{(P_{A2} - P_{A1})}{x_2 - x_1}$$

A similar equation may be derived for the counterdiffusion of gas B.

REFERENCES

1. Bard, Allen J.; Faulkner, Larry R. (January 2001). Electrochemical methods: fundamentals and applications. New York: Wiley. ISBN 978-0-471-04372-0. Retrieved 27 February 2009.
2. Geiger, William E. (2007-11-01). "Organometallic Electrochemistry: Origins, Development, and Future". Organometallics. 26 (24): 5738–5765. doi:10.1021/om700558k.
3. The Supplement (1803 edition) to Encyclopedia Britannica 3rd edition (1797), volume 1, page 225, "Mister Van Marum, by means of his great electrical machine, decomposed the calces of tin, zinc, and antimony, and resolved them into their respective metals and oxygen" and gives as a reference Journal de Physiques, 1785.
4. Crookes, William (1875). The Chemical news and journal of industrial science; with which is incorporated the "Chemical gazette.": A journal of practical chemistry in all its applications to pharmacy, arts and manufactures. Chemical news office. pp. 294–. Retrieved 27 February 2011.
5. Tilley, R. J. D. (2004). Understanding solids: the science of materials. John Wiley and Sons. pp. 281–. ISBN 978-0-470-85276-7. Retrieved 22 October 2011.
6. "Rust Removal using Electrolysis". antique-engines.com. Retrieved April 1, 2015.
7. Atkins, Peter (1997). Physical Chemistry, 6th edition (W.H. Freeman and Company, New York).
8. Vanýsek, Petr (2007). "Electrochemical Series", in Handbook of Chemistry and Physics: 88th Edition (Chemical Rubber Company).
9. Vogel, A.E. (1951), A textbook of Quantitative Inorganic Analysis, Longmans,

Green and Co

10. Lee, Seunghwa; Ju, Hyungkuk; Machunda, Revocatus; Uhm, Sunghyun; Lee, Jae Kwang; Lee, Hye Jin; Lee, Jaeyoung (2015). "Sustainable production of formic acid by electrolytic reduction of gaseous carbon dioxide". J. Mater. Chem. A. 3 (6): 3029. doi:10.1039/C4TA03893B.
11. Whipple, Devin T.; Kenis, Paul J. A. (2010). "Prospects of CO2 Utilization via Direct Heterogeneous Electrochemical Reduction". The Journal of Physical Chemistry Letters. 1 (24): 3451. doi:10.1021/jz1012627.
12. Machunda, Revocatus L.; Ju, Hyungkuk; Lee, Jaeyoung (2011). "Electrocatalytic reduction of CO2 gas at Sn based gas diffusion electrode". Current Applied Physics. 11 (4): 986. doi:10.1016/j.cap.2011.01.003.
13. Hori, Y. "Electrochemical CO2 Reduction on Metal Electrodes" (2008) pp. 141–153 in Modern Aspects of Electrochemistry, ed. C. G. Vayeanas, R. White and M. E. Gamboa-Aldeco, Springer, New York, no. 42. doi:10.1007/978-0-387-49489-0_3 ISBN 978-0-387-49489-0.
14. Appel, Aaron M.; Bercaw, John E.; Bocarsly, Andrew B.; Dobbek, Holger; Dubois, Daniel L.; Dupuis, Michel; Ferry, James G.; Fujita, Etsuko; Hille, Russ; Kenis, Paul J. A.; Kerfeld, Cheryl A.; Morris, Robert H.; Peden, Charles H. F.; Portis, Archie R.; Ragsdale, Stephen W.; Rauchfuss, Thomas B.; Reek, Joost N. H.; Seefeldt, Lance C.; Thauer, Rudolf K.; Waldrop, Grover L. (2013). "Frontiers, Opportunities, and Challenges in Biochemical and Chemical Catalysis of CO2 Fixation". Chemical Reviews. 113 (8): 6621. doi:10.1021/cr300463y. PMID 23767781.
15. Qiao, Jinli; Liu, Yuyu; Hong, Feng; Zhang, Jiujun (2014). "A review of catalysts for the electroreduction of carbon dioxide to produce low-carbon fuels". Chem. Soc. Rev. 43 (2): 631. doi:10.1039/C3CS60323G. PMID 24186433.
16. Carmo, M; Fritz D; Mergel J; Stolten D (2013). "A comprehensive review on PEM water electrolysis". Journal of Hydrogen Energy. doi:10.1016/j.ijhydene.2013.01.151.
17. Zittel, Werner; Wurster, Reinhold (8 July 1996). "Chapter 3: Production of Hydrogen. Part 4: Production from electricity by means of electrolysis". HyWeb: Knowledge – Hydrogen in the Energy Sector. Ludwig-Bölkow-Systemtechnik GmbH. Archived from the original on 7 February 2007.
18. Levene, J.; Kroposki, B.; Sverdrup, G. (March 2006). "Wind Energy and Production of Hydrogen and Electricity – Opportunities for Renewable Hydrogen – Preprint" (PDF). National Renewable Energy Laboratory. Retrieved 20 October 2008.
19. Ju, Hyungkuk; Giddey, Sarbjit; Badwal, Sukhvinder P.S.; Mulder, Roger J. (2016). "Electro-catalytic conversion of ethanol in solid electrolyte cells for distributed hydrogen generation". Electrochimica Acta. 212: 744. doi:10.1016/j.electacta.2016.07.062.
20. Giddey, S.; Kulkarni, A.; Badwal, S.P.S. (2015). "Low emission hydrogen generation through carbon assisted electrolysis". International Journal of Hydrogen

Energy. 40: 70. doi:10.1016/j.ijhydene.2014.11.033.

21. "Superconductivity in an organic solid. Synthesis, structure, and conductivity of bis(tetramethyltetraselenafulvalenium) perchlorate, (TMTSF)2ClO4". Journal of the American Chemical Society. 103 (9): 2440. 1981. doi:10.1021/ja00399a065.
22. Williams, Jack M (2007). "Highly Conducting and Superconducting Synthetic Metals". Inorganic Syntheses. 26: 386–394. doi:10.1002/9780470132579.ch70.
23. Kittel, Charles. Introduction to Solid State Physics, 7th Edition. Wiley.
24. Riess, What does a voltmeter measure? Solid State Ionics 95, 327 (1197)
25. Sah, Chih-Tang (1991). Fundamentals of Solid-State Electronics. World Scientific. p. 404. ISBN 9810206372.
26. Datta, Supriyo (2005). Quantum Transport: Atom to Transistor. Cambridge University Presss. p. 7. ISBN 9780521631457.
27. Kittel, Charles; Herbert Kroemer (1980-01-15). Thermal Physics (2nd Edition). W. H. Freeman. p. 357. ISBN 978-0-7167-1088-2.
28. Sze, S. M. (1964). Physics of Semiconductor Devices. Wiley. ISBN 0-471-05661-8.
29. Sommerfeld, Arnold (1964). Thermodynamics and Statistical Mechanics. Academic Press.
30. "3D Fermi Surface Site". Phys.ufl.edu. 1998-05-27. Retrieved 2013-04-22.
31. For example: D. Chattopadhyay (2006). Electronics (fundamentals And Applications). ISBN 978-81-224-1780-7. and Balkanski and Wallis (2000-09-01). Semiconductor Physics and Applications. ISBN 978-0-19-851740-5.
32. Technically, it is possible to consider the vacuum to be an insulator and in fact its Fermi level is defined if its surroundings are in equilibrium. Typically however the Fermi level is two to five electron volts below the vacuum electrostatic potential energy, depending on the work function of the nearby vacuum wall material. Only at high temperatures will the equilibrium vacuum be populated with a significant number of electrons (this is the basis of thermionic emission).
33. Shegelski, Mark R. A. (May 2004). "The chemical potential of an ideal intrinsic semiconductor". American Journal of Physics. 72 (5): 676–678. Bibcode:2004AmJPh..72..676S. doi:10.1119/1.1629090.
34. Beenakker, C. W. J. (1991). "Theory of Coulomb-blockade oscillations in the conductance of a quantum dot". Physical Review B. 44 (4): 1646. Bibcode:1991PhRvB..44.1646B. doi:10.1103/PhysRevB.44.1646.
35. Bard; Faulkner. "Section 2.2.4(a),4-5". Electrochemical Methods (2nd ed.).
36. Madelung, Otfried. Introduction to solid-state theory. p. 198.
37. Ashcroft; Mermin. Solid State Physics. p. 593.
38. IUPAC, Compendium of Chemical Terminology, 2nd ed. (the "Gold Book") (1997). Online corrected version: (2006–) "Electrochemical potential".
39. Grover, D.J. (1996). Modeling water chemistry and electrochemical corrosion potential in boiling water reactors (PDF) (Thesis). Massachusetts Institute of

Technology.

40. Nelson, David; Cox, Michael (2013). Lehninger Principles of Biochemistry. New York: W.H. Freeman. p. 403. ISBN 978-1-4292-3414-6.
41. Nath, Sunil; Villadsen, John (2015-03-01). "Oxidative phosphorylation revisited". Biotechnology and Bioengineering. 112 (3): 429–437. doi:10.1002/bit.25492. ISSN 1097-0290.
42. Yang, Huanghe; Zhang, Guohui; Cui, Jianmin (2015-01-01). "BK channels: multiple sensors, one activation gate". Membrane Physiology and Membrane Biophysics. 6: 29. doi:10.3389/fphys.2015.00029. PMC 4319557Freely accessible. PMID 25705194.
43. Shattock, Michael J.; Ottolia, Michela; Bers, Donald M.; Blaustein, Mordecai P.; Boguslavskyi, Andrii; Bossuyt, Julie; Bridge, John H. B.; Chen-Izu, Ye; Clancy, Colleen E. (2015-03-15). "Na+/Ca2+ exchange and Na+/K+-ATPase in the heart". The Journal of Physiology. 593 (6): 1361–1382. doi:10.1113/jphysiol.2014.282319. ISSN 1469-7793. PMC 4376416Freely accessible. PMID 25772291.
44. Aperia, Anita; Akkuratov, Evgeny E.; Fontana, Jacopo Maria; Brismar, Hjalmar (2016-04-01). "Na+-K+-ATPase, a new class of plasma membrane receptors". American Journal of Physiology - Cell Physiology. 310 (7): C491–C495. doi:10.1152/ajpcell.00359.2015. ISSN 0363-6143. PMID 26791490.
45. Nelson, David; Cox, Michael (2013). Lehninger Principles of Biochemistry. New York: W.H. Freeman. p. 464. ISBN 978-1-4292-3414-6.
46. Poburko, Damon; Demaurex, Nicolas (2012-04-24). "Regulation of the mitochondrial proton gradient by cytosolic Ca2+ signals". Pflügers Archiv - European Journal of Physiology. 464 (1): 19–26. doi:10.1007/s00424-012-1106-y. ISSN 0031-6768.
47. Nelson, David; Cox, Michael (2013). Lehninger Principles of Biochemistry. New York: W.H. Freeman. pp. 743–745. ISBN 978-1-4292-3414-6.
48. Nelson, David; Cox, Michael (2013). Lehninger Principles of Biochemistry. New York: W.H. Freeman. p. 744. ISBN 978-1-4292-3414-6.
49. Nelson, David; Cox, Michael (2013). Lehninger Principles of Biochemistry. New York: W.H. Freeman. pp. 769–770. ISBN 978-1-4292-3414-6.
50. Nelson, David; Cox, Michael (2013). Lehninger Principles of Biochemistry. New York: W.H. Freeman. p. 770. ISBN 978-1-4292-3414-6.
51. Höhner, Ricarda; Aboukila, Ali; Kunz, Hans-Henning; Venema, Kees (2016-01-01). "Proton Gradients and Proton-Dependent Transport Processes in the Chloroplast". Plant Physiology. 7: 218. doi:10.3389/fpls.2016.00218. PMC 4770017Freely accessible. PMID 26973667.
52. Nelson, David; Cox, Michael (2013). Lehninger Principles of Biochemistry. New York: W.H. Freeman. pp. 464–465. ISBN 978-1-4292-3414-6.
53. Eisenberg, Bob (2013-05-07). "Interacting Ions in Biophysics: Real is not Ideal". Biophysical Journal. 104 (9): 1849–1866. doi:10.1016/j.bpj.2013.03.049. PMC

3647150Freely accessible. PMID 23663828.

54. Nelson, David; Cox, Michael (2013). Lehninger Principles of Biochemistry. New York: W.H. Freeman. p. 465. ISBN 978-1-4292-3414-6.
55. Gunner, M. R.; Amin, Muhamed; Zhu, Xuyu; Lu, Jianxun (2013-08-01). "Molecular mechanisms for generating transmembrane proton gradients". Biochimica et Biophysica Acta (BBA) - Bioenergetics. Metals in Bioenergetics and Biomimetics Systems. 1827 (8–9): 892–913. doi:10.1016/j.bbabio.2013.03.001. PMC 3714358Freely accessible. PMID 23507617.
56. Wickstrand, Cecilia; Dods, Robert; Royant, Antoine; Neutze, Richard (2015-03-01). "Bacteriorhodopsin: Would the real structural intermediates please stand up?". Biochimica et Biophysica Acta (BBA) - General Subjects. Structural biochemistry and biophysics of membrane proteins. 1850 (3): 536–553. doi:10.1016/j.bbagen.2014.05.021.
57. Nelson, David; Cox, Michael (2013). Lehninger Principles of Biochemistry. New York: W.H. Freeman. pp. 782–783. ISBN 978-1-4292-3414-6.
58. Schöttler, Mark Aurel; Tóth, Szilvia Z.; Boulouis, Alix; Kahlau, Sabine (2015-05-01). "Photosynthetic complex stoichiometry dynamics in higher plants: biogenesis, function, and turnover of ATP synthase and the cytochrome b 6 f complex". Journal of Experimental Botany. 66 (9): 2373–2400. doi:10.1093/jxb/eru495. ISSN 0022-0957. PMID 25540437.
59. Sun, Fei; Zhou, Qiangjun; Pang, Xiaoyun; Xu, Yingzhi; Rao, Zihe (2013-08-01). "Revealing various coupling of electron transfer and proton pumping in mitochondrial respiratory chain". Current Opinion in Structural Biology. 23 (4): 526–538. doi:10.1016/j.sbi.2013.06.013.

CHAPTER 8

Electrochemical Cell

Chapter Contents

- **Introduction**
- **Half-Cells**
- **Cell Notation**
- **Electrochemical Machining**
- **Electro-Mechanical Modeling**
- **Electrical Discharge Machining**

An extremely important class of oxidation and reduction reactions are used to provide useful electrical energy in batteries. A simple electrochemical cell can be made from copper and zinc metals with solutions of their sulfates. In the process of the reaction, electrons can be transferred from the zinc to the copper through an electrically conducting path as a useful electric current.

An electrochemical cell can be created by placing metallic electrodes into an electrolyte where a chemical reaction either uses or generates an electric current. Electrochemical cells which generate an electric current are called voltaic cells or galvanic cells, and common batteries consist of one or more such cells. In other electrochemical cells an externally supplied electric current is used to drive a chemical reaction which would not occur spontaneously. Such cells are called electrolytic cells.

INTRODUCTION

An electrochemical cell is a device capable of either generating electrical energy from chemical reactions or facilitating chemical reactions through the introduction of electrical energy. A common example of an electrochemical cell is a standard 1.5-volt cell meant for consumer use. This type of device is known as a single Galvanic cell. A *battery* consists of two or more cells, connected in either parallel or series pattern.

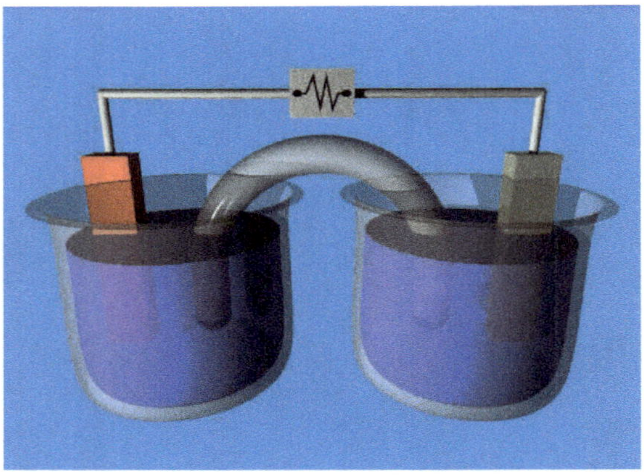

A demonstration electrochemical cell setup resembling the Daniell cell. The two half-cells are linked by a salt bridge carrying ions between them. Electrons flow in the external circuit.

HALF-CELLS

An electrochemical cell consists of two half-cells. Each *half-cell* consists of an electrode and an electrolyte. The two half-cells may use the same electrolyte, or they may use different electrolytes. The chemical reactions in the cell may involve the electrolyte, the electrodes, or an external substance (as in fuel cells that may use hydrogen gas as a reactant). In a full electrochemical cell, species from one half-cell lose electrons (oxidation) to their electrode while species from the other half-cell gain electrons (reduction) from their electrode.

A *Salt Bridge* (e.g., filter paper soaked in KNO_3, NaCl, or some other electrolyte) is often employed to provide ionic contact between two half-cells with different electrolytes, yet prevent the solutions from mixing and causing unwanted side reactions. An alternative to a salt bridge is to allow direct contact (and mixing) between the two half-cells, for example in simple electrolysis of water.

As electrons flow from one half-cell to the other through an external circuit, a difference in charge is established. If no ionic contact were provided, this charge difference would quickly prevent the further flow of electrons. A salt bridge allows the flow of negative or positive ions to maintain a steady-state charge distribution between the oxidation and reduction vessels, while keeping the contents otherwise separate. Other devices for achieving separation of solutions are porous pots and gelled solutions. A porous pot is used in the Bunsen cell (right).

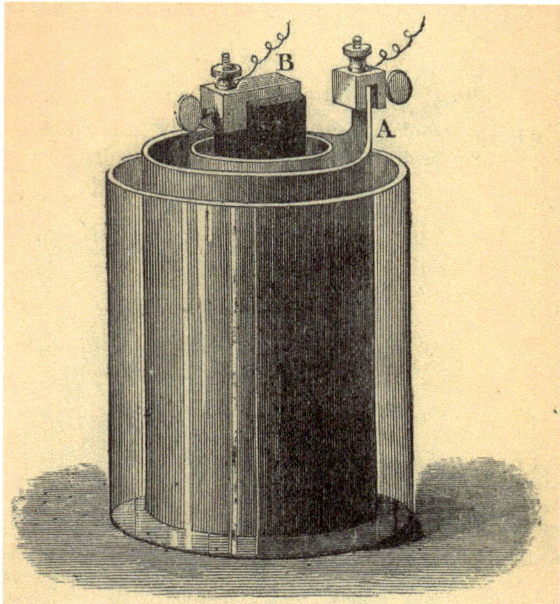

The Bunsen cell, invented by Robert Bunsen.

Equilibrium Reaction

Each half-cell has a characteristic voltage. Various choices of substances for each half-cell give different potential differences. Each reaction is undergoing an equilibrium reaction between different oxidation states of the ions: When equilibrium is reached, the cell cannot provide further voltage. In the half-cell that is undergoing oxidation, the closer the equilibrium lies to the ion/atom with the more positive oxidation state the more potential this reaction will provide. Likewise, in the reduction reaction, the closer the equilibrium lies to the ion/atom with the more *negative* oxidation state the higher the potential.

Cell Potential

The cell potential can be predicted through the use of electrode potentials (the voltages of each half-cell). These half-cell potentials are defined relative to the assignment of 0 volts to the standard hydrogen electrode (SHE). The difference in voltage between electrode potentials gives a prediction for the potential measured. When calculating the difference in voltage, one must first rewrite the half-cell reaction equations to obtain a balanced oxidation-reduction equation.

- Reverse the reduction reaction with the smallest potential (to create an oxidation reaction/ overall positive cell potential)
- Half-reactions must be multiplied by integers to achieve electron balance.

Note that the cell potential does not change when the reaction is multiplied by a constant.

Cell potentials have a possible range of roughly zero to 6 volts. Cells using water-based electrolytes are usually limited to cell potentials less than about 2.5 volts, because the very powerful oxidizing and reducing agents that would be required to produce a higher cell potential tend to react with the water. Higher cell potentials are possible with cells using other solvents instead of water. For instance, lithium cells with a voltage of 3 volts are commonly available.

The cell potential depends on the concentration of the reactants, as well as their type. As the cell is discharged, the concentration of the reactants decreases, and the cell potential also decreases.

CELL NOTATION

Cell notation in chemistry is a shorthand way of expressing a certain reaction in an electrochemical cell. The cell anode and cathode (half-cells) are separated by two bars or slashes representing a salt bridge, with the anode on the left and cathode on the right. Individual solid, liquid or aqueous phases within each half-cell are separated by a single bar. Concentrations of dissolved species, in each phase written in parentheses and the state of each phase (usually s (solid), l (liquid), g (gas) or aq. (aqueous solution)) is included in a subscript after the species name.

Some examples of this notation are:

$Zn^{\circ}|Zn^{2+},Cl^{-}|AgCl|Ag^{\circ}$. This means that the left electrode (anode) is made of zinc, while the other one (right, cathode) is composed of a silver wire covered by a silver chloride layer which is not soluble. Both of the electrodes are immersed into aqueous media where are present zinc and chloride ions.

$Zn^{\circ}|Zn^{2+}$, $SO_4^{2-}||SO_4^{2-}|Cu^{2+}|Cu^{\circ}$. This cell is very famous: The Daniell cell. If the electrodes are connected, a spontaneous reaction takes place. Zinc is oxidized, and copper ions are reduced.

Sometimes the state of each species into the cell is written. For example, in the zinc cell (showed above), we can write that zinc, silver and silver chloride are solids, while zinc cation and chloride anion are in aqueous media. So, the new notation will be:

$Zn^{\circ}(s)|Zn^{2+}(aq),Cl^{-}(aq)|AgCl(s)|Ag^{\circ}(s)$

It is possible to express the ion concentration too. For example, in the Daniell cell:

$Zn^{\circ}(s)|Zn^{2+}$(aq, 1 mol/L), SO_4^{2-}(aq, 1 mol/L)$||SO_4^{2-}$(aq, 1 mol/L)$|Cu^{2+}$(aq, 1 mol/L)$|Cu^{\circ}(s)$

In this case, all ions (sulfate, zinc and copper) are in a concentration equal to 1 mol/L.

ELECTROCHEMICAL MACHINING

Electrochemical machining (ECM) is a method of removing metal by an electrochemical process. It is normally used for mass production and is used for working extremely hard materials or materials that are difficult to machine using conventional methods. Its use is limited to electrically conductive materials. ECM can cut small or odd-shaped angles, intricate contours or cavities in hard and exotic metals, such as titanium aluminides, Inconel, Waspaloy, and high nickel, cobalt, and rhenium alloys. Both external and internal geometries can be machined.

ECM is often characterized as "reverse electroplating", in that it removes material instead of adding it. It is similar in concept to electrical discharge machining (EDM) in that a high current is passed between an electrode and the part, through an electrolytic material removal process having a negatively charged electrode (cathode), a conductive fluid (electrolyte), and a conductive workpiece (anode); however, in ECM there is no tool wear. The ECM cutting tool is guided along the desired path close to the work but without touching the piece. Unlike EDM, however, no sparks are created. High metal removal rates are possible with ECM, with no thermal or mechanical stresses being transferred to the part, and mirror surface finishes can be achieved.

In the ECM process, a cathode (tool) is advanced into an anode (workpiece). The pressurized electrolyte is injected at a set temperature to the area being cut. The feed rate is the same as the rate of "liquefication" of the material. The gap between the tool and the workpiece varies within 80–800 micrometers (0.003–0.030 in.) As electrons cross the gap, material from the workpiece is dissolved, as the tool forms the desired shape in the workpiece. The electrolytic fluid carries away the metal hydroxide formed in the process.

As far back as 1929, an experimental ECM process was developed by W.Gussef, although it was 1959 before a commercial process was established by the Anocut Engineering Company. B.R. and J.I. Lazarenko are also credited with proposing the use of electrolysis for metal removal.

Much research was done in the 1960s and 1970s, particularly in the gas turbine industry. The rise of EDM in the same period slowed ECM research in the west, although work continued behind the Iron Curtain. The original problems of poor dimensional accuracy and environmentally polluting waste have largely been overcome, although the process remains a niche technique.

The ECM process is most widely used to produce complicated shapes such as turbine blades with good surface finish in difficult to machine materials. It is also widely and effectively used as a deburring process.

In deburring, ECM removes metal projections left from the machining process, and so dulls sharp edges. This process is fast and often more convenient than the conventional methods of deburring by hand or nontraditional machining processes.

Advantages

- Complex, concave curvature components can be produced easily by using convex and concave tools.
- Tool wear is zero, same tool can be used for producing infinite number of components.
- No direct contact between tool and work material so there are no forces and residual stresses.

- The surface finish produced is excellent.
- Less amount of heat is generated.

Disadvantages

The saline (or acidic) electrolyte poses the risk of corrosion to tool, workpiece and equipment.

Only electrically conductive materials can be machined. High Specific Energy consumption.

Currents Involved

The needed current is proportional to the desired rate of material removal, and the removal rate in mm/minute is proportional to the amps per square mm.

Typical currents range from 0.1 amp per square mm to 5 amps per square mm. Thus, for a small plunge cut of a 1 by 1 mm tool with a slow cut, only 0.1 amps would be needed.

However, for a higher feed rate over a larger area, more current would be used, just like any machining process—removing more material faster takes more power.

Thus, if a current density of 4 amps per square millimeter was desired over a 100×100 mm area, it would take 40,000 amps (and lots of coolant/electrolyte).

Setup and Equipment

An ET 3000 ECM machine by INDEC of Russia.

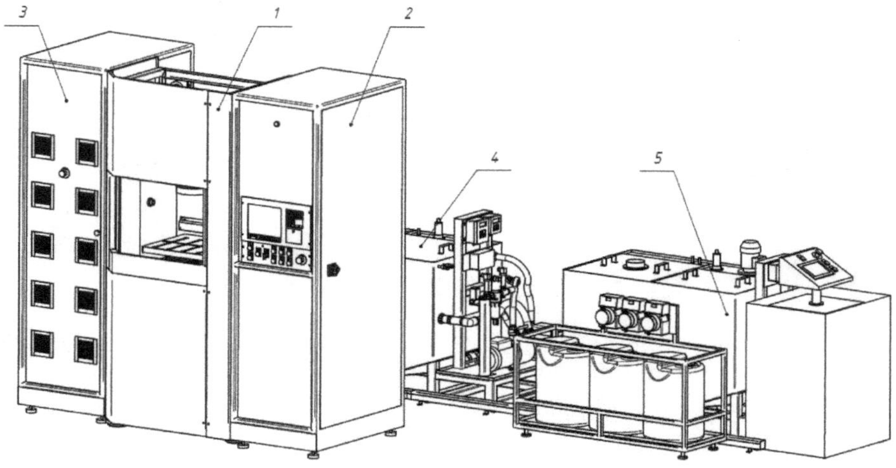

ECM machines come in both vertical and horizontal types. Depending on the work requirements, these machines are built in many different sizes as well. The vertical machine consists of a base, column, table, and spindle head. The spindle head has a servo-mechanism that automatically advances the tool and controls the gap between the cathode (tool) and the workpiece.

CNC Machines of Up to Six Axes Are Available.

Copper is often used as the electrode material. Brass, graphite, and copper-tungsten are also often used because they are easily machined, they are conductive materials, and they will not corrode.

Applications

Some of the very basic applications of ECM include:

- Die-sinking operations
- Drilling jet engine turbine blades
- Multiple hole drilling
- Machining steam turbine blades within close limits

Similarities between EDM and ECM

- The tool and workpiece are separated by a very small gap, i.e. no contact in between them is made.
- The tool and material must both be conductors of electricity.
- Needs high capital investment.

- Systems consume lots of power.
- A fluid is used as a medium between the tool and the work piece (conductive for ECM and dielectric for EDM).
- The tool is fed continuously towards the workpiece to maintain a constant gap between them (ECM may incorporate intermittent or cyclic, typically partial, tool withdrawal).

Difference between ECM and ECG

- Electrochemical grinding (ECG) is similar to electrochemical machining (ECM) but uses a contoured conductive grinding wheel instead of a tool shaped like the contour of the workpiece.

ELECTRO-MECHANICAL MODELING

The purpose of Electro-Mechanical Modeling is to model and simulate an electro-mechanical system, such that it's physical parameters can be examined before the actual system is built. Parameter estimation and physical realization of the overall system is the major design objective of Electro-Mechanical modeling. Theory driven mathematical model can be used or applied to other system to judge the performance of the joint system as a whole. This is a well known & proven technique for designing large control system for industrial as well as academic multi-disciplinary complex system.

Different Types of Mathematical Modelling

The modeling of purely mechanical systems is mainly based on the Lagrangian which is a function of the generalized coordinates and the associated velocities. If all forces are derivable from a potential, then the time behavior of the dynamical systems is completely determined. For simple mechanical systems, the Lagrangian is defined as the difference of the kinetic energy and the potential energy.

There exists a similar approach for electrical system. By means of the electrical co-energy and well defined power quantities, the equations of motions are uniquely defined. The currents of the inductors and the voltage drops across the capacitors play the role of the generalized coordinates. All constraints, for instance caused by the Kirchhoff laws, are eliminated from the considerations. After that, a suitable transfer function is to be derived from the system parameters which eventually governs the behavior of the system.

In consequence, we have quantities (kinetic and potential energy, generalized forces) which determine the mechanical part and quantities (co-energy, powers) for the description of the electrical part. This offers a combination of the mechanical and electrical parts by means of an energy approach. As a result, an extended Lagrangian format is produced.

ELECTRICAL DISCHARGE MACHINING

Electrical discharge machining (EDM), also known as spark machining, spark eroding, burning, die sinking, wire burning or wire erosion, is a manufacturing process whereby a desired shape is obtained by using electrical discharges (sparks). Material is removed from the workpiece by a series of rapidly recurring current discharges between two electrodes, separated by a dielectric liquid and subject to an electric voltage. One of the electrodes is called the tool-electrode, or simply the "tool" or "electrode," while the other is called the workpiece-electrode, or "workpiece." The process depends upon the tool and workpiece not making actual contact.

An electrical discharge machine.

When the voltage between the two electrodes is increased, the intensity of the electric field in the volume between the electrodes becomes greater than the strength of the dielectric (at least in some places), which breaks down, allowing current to flow between the two electrodes. This phenomenon is the same as the breakdown of a capacitor (condenser) . As a result, material is removed from the electrodes. Once the current stops (or is stopped, depending on the type of generator), new liquid dielectric is usually conveyed into the inter-electrode volume, enabling the solid particles (debris) to be carried away and the insulating properties of the dielectric to be restored. Adding new liquid dielectric in the inter-electrode volume is commonly referred to as "flushing." Also, after a current flow, the difference of potential between the electrodes is restored to what it was before the breakdown, so that a new liquid dielectric breakdown can occur.

History

The erosive effect of electrical discharges was first noted in 1770 by English physicist Joseph Priestley.

Die-sink EDM

Two Russian scientists, B. R. Butinzky and N. I. Lazarenko, were tasked in 1943 to investigate ways of preventing the erosion of tungsten electrical contacts due to sparking. They failed in this task but found that the erosion was more precisely controlled if the electrodes were immersed in a dielectric fluid. This led them to invent an EDM machine used for working difficult-to-machine materials such as tungsten. The Lazarenkos' machine is known as an R-C-type machine, after the resistor–capacitor circuit (RC circuit) used to charge the electrodes.

Simultaneously but independently, an American team, Harold Stark, Victor Harding, and Jack Beaver, developed an EDM machine for removing broken drills and taps from aluminum castings. Initially constructing their machines from feeble electric-etching tools, they were not very successful. But more powerful sparking units, combined with automatic spark repetition and fluid replacement with an electromagnetic interrupter arrangement produced practical machines. Stark, Harding, and Beaver's machines were able to produce 60 sparks per second. Later machines based on their design used vacuum tube circuits that were able to produce thousands of sparks per second, significantly increasing the speed of cutting.

Wire-cut EDM

The wire-cut type of machine arose in the 1960s for the purpose of making tools (dies) from hardened steel. The tool electrode in wire EDM is simply a wire. To avoid the erosion of material from the wire causing it to break, the wire is wound between two spools so that the active part of the wire is constantly changing. The earliest numerical controlled (NC) machines were conversions of punched-tape vertical milling machines. The first commercially available NC machine built as a wire-cut EDM machine was manufactured in the USSR in 1967. Machines that could optically follow lines on a master drawing were developed by David H. Dulebohn's group in the 1960s at Andrew Engineering Company for milling and grinding machines. Master drawings were later produced by computer numerical controlled (CNC) plotters for greater accuracy. A wire-cut EDM machine using the CNC drawing plotter and optical line follower techniques was produced in 1974. Dulebohn later used the same plotter CNC program to directly control the EDM machine, and the first CNC EDM machine was produced in 1976.

Generalities

Electrical discharge machining is a machining method primarily used for hard metals or those that would be very difficult to machine with traditional techniques. EDM typically works with materials that are electrically conductive, although methods for machining insulating ceramics with EDM have also been proposed. EDM can cut intricate contours or cavities in pre-hardened steel without the need for heat treatment to soften and re-harden them. This method can be used with any other metal or metal alloy such as titanium, hastelloy, kovar, and inconel. Also, applications of this process to shape polycrystalline diamond tools have been reported.

EDM is often included in the "non-traditional" or "non-conventional" group of machining methods together with processes such as electrochemical machining (ECM), water jet cutting (WJ, AWJ), laser cutting and opposite to the "conventional" group (turning, milling, grinding, drilling and any other process whose material removal mechanism is essentially based on mechanical forces).

Ideally, EDM can be seen as a series of breakdown and restoration of the liquid dielectric in-between the electrodes. However, caution should be exerted in considering such a statement because it is an idealized model of the process, introduced to describe the fundamental ideas underlying the process. Yet, any practical application involves many aspects that may also need to be considered. For instance, the removal of the debris from the inter-electrode volume is likely to be always partial. Thus the electrical properties of the dielectric in the inter-electrodes volume can be different from their nominal values and can even vary with time. The inter-electrode distance, often also referred to as spark-gap, is the end result of the control algorithms of the specific machine used. The control of such a distance appears logically to be central to this process. Also, not all of the current between the dielectric is of the ideal type described above: the spark-gap can be short-circuited by the debris. The control system of the electrode may fail to react quickly enough to prevent the two electrodes (tool and workpiece) from coming into contact, with a consequent short circuit. This is unwanted because a short circuit contributes to material removal differently from the ideal case. The flushing action can be inadequate to restore the insulating properties of the dielectric so that the current always happens in the point of the inter-electrode volume (this is referred to as arcing), with a consequent unwanted change of shape (damage) of the tool-electrode and workpiece. Ultimately, a description of this process in a suitable way for the specific purpose at hand is what makes the EDM area such a rich field for further investigation and research.

To obtain a specific geometry, the EDM tool is guided along the desired path very close to the work; ideally it should not touch the workpiece, although in reality this may happen due to the performance of the specific motion control in use. In this way, a large number of current discharges (colloquially also called sparks) happen, each contributing to the removal of material from both tool and workpiece, where small craters are formed. The size of the craters is a function of the technological parameters set for the specific job at hand. They can be with typical dimensions ranging from the nanoscale (in micro-EDM operations) to some hundreds of micrometers in roughing conditions.

The presence of these small craters on the tool results in the gradual erosion of the electrode. This erosion of the tool-electrode is also referred to as wear. Strategies are needed to counteract the detrimental effect of the wear on the geometry of the workpiece. One possibility is that of continuously replacing the tool-electrode during a machining operation. This is what happens if a continuously replaced wire is used as electrode. In this case, the correspondent EDM process is also called wire EDM. The tool-electrode can also be used in such a way that only a small portion of it is actually engaged in the machining process and this portion is changed on a regular basis. This is, for instance, the case when using a rotating disk as a tool-electrode. The corresponding process is often also referred to as EDM grinding.

A further strategy consists in using a set of electrodes with different sizes and shapes during the same EDM operation. This is often referred to as multiple electrode strategy, and is most common when the tool electrode replicates in negative the wanted shape and is advanced towards the blank along a single direction, usually the vertical direction (i.e. z-axis). This resembles the sink of the tool into the dielectric liquid in which the workpiece is immersed, so, not surprisingly, it is often referred to as die-sinking EDM (also called conventional EDM and ram EDM). The corresponding machines are often called sinker EDM. Usually, the electrodes of this type have quite complex forms. If the final geometry is obtained using a usually simple-shaped electrode which is moved along several directions and is possibly also subject to rotations, often the term EDM milling is used.

In any case, the severity of the wear is strictly dependent on the technological parameters used in the operation (for instance: polarity, maximum current, open circuit voltage). For example, in micro-EDM, also known as μ-EDM, these parameters are usually set at values which generates severe wear. Therefore, wear is a major problem in that area.

The problem of wear to graphite electrodes is being addressed. In one approach, a digital generator, controllable within milliseconds, reverses polarity as electro-erosion takes place. That produces an effect similar to electroplating that continuously deposits the eroded graphite back on the electrode. In another method, a so-called "Zero Wear" circuit reduces how often the discharge starts and stops, keeping it on for as long a time as possible.

Definition of the Technological Parameters

Difficulties have been encountered in the definition of the technological parameters that drive the process.

Two broad categories of generators, also known as power supplies, are in use on EDM machines commercially available: the group based on RC circuits and the group based on transistor controlled pulses.

In the first category, the main parameters to choose from at setup time are the resistance(s) of the resistor(s) and the capacitance(s) of the capacitor(s). In an ideal condition these quantities would affect the maximum current delivered in a discharge which is expected to be associated with the charge accumulated on the capacitors at a certain moment in time. Little control, however, is expected over the time duration of the discharge, which is likely to depend on the actual spark-gap conditions (size and pollution) at the moment of the discharge. The RC circuit generator can allow the user to obtain short time durations of the discharges more easily than the pulse-controlled generator, although this advantage is diminishing with the development of new electronic components. Also, the open circuit voltage (i.e. the voltage between the electrodes when the dielectric is not yet broken) can be identified as steady state voltage of the RC circuit.

In generators based on transistor control, the user is usually able to deliver a train of pulses of voltage to the electrodes. Each pulse can be controlled in shape, for instance, quasi-rectangular. In particular, the time between two consecutive pulses and the duration of each pulse can be set. The amplitude of each pulse constitutes the open circuit voltage. Thus, the maximum duration of discharge is equal to the duration of a pulse of voltage in the train. Two pulses of current are then expected not to occur for a duration equal or larger than the time interval between two consecutive pulses of voltage.

The maximum current during a discharge that the generator delivers can also be controlled. Because other sorts of generators may also be used by different machine builders, the parameters that may actually be set on a particular machine will depend on the generator manufacturer. The details of the generators and control systems on their machines are not always easily available to their user. This is a barrier to describing unequivocally the technological parameters of the EDM process. Moreover, the parameters affecting the phenomena occurring between tool and electrode are also related to the controller of the motion of the electrodes.

A framework to define and measure the electrical parameters during an EDM operation directly on inter-electrode volume with an oscilloscope external to the machine has been recently proposed by Ferri *et al.* These authors conducted their research in the field of µ-EDM, but the same approach can be used in any EDM operation. This would enable the user to estimate directly the electrical parameters that affect their operations without relying upon machine manufacturer›s claims. When machining different materials in the same setup conditions, the actual electrical parameters of the process are significantly different.

Material Removal Mechanism

The first serious attempt of providing a physical explanation of the material removal during electric discharge machining is perhaps that of Van Dijck. Van Dijck presented a thermal model together with a computational simulation to explain the phenomena between the electrodes during electric discharge machining. However, as Van Dijck himself admitted in his study, the number of assumptions made to overcome the lack of experimental data at that time was quite significant.

Further models of what occurs during electric discharge machining in terms of heat transfer were developed in the late eighties and early nineties, including an investigation at Texas A&M University with the support of AGIE, now Agiecharmilles. It resulted in three scholarly papers: the first presenting a thermal model of material removal on the cathode, the second presenting a thermal model for the erosion occurring on the anode and the third introducing a model describing the plasma channel formed during the passage of the discharge current through the dielectric liquid. Validation of these models is supported by experimental data provided by AGIE.

These models give the most authoritative support for the claim that EDM is a thermal process, removing material from the two electrodes because of melting and/or vaporization, along with pressure dynamics established in the spark-gap by the collapsing of the plasma channel. However, for small discharge energies the models are inadequate to explain the experimental data. All these models hinge on a number of assumptions from such disparate research areas as submarine explosions, discharges in gases, and failure of transformers, so it is not surprising that alternative models have been proposed more recently in the literature trying to explain the EDM process.

Among these, the model from Singh and Ghosh reconnects the removal of material from the electrode to the presence of an electrical force on the surface of the electrode that could mechanically remove material and create the craters. This would be possible because the material on the surface has altered mechanical properties due to an increased temperature caused by the passage of electric current. The authors' simulations showed how they might explain EDM better than a thermal model (melting and/or evaporation), especially for small discharge energies, which are typically used in μ-EDM and in finishing operations.

Given the many available models, it appears that the material removal mechanism in EDM is not yet well understood and that further investigation is necessary to clarify it, especially considering the lack of experimental scientific evidence to build and validate the current EDM models. This explains an increased current research effort in related experimental techniques.

Types

Sinker EDM

Sinker EDM, also called cavity type EDM or volume EDM, consists of an electrode and workpiece submerged in an insulating liquid such as, more typically, oil or, less frequently, other dielectric fluids. The electrode and workpiece are connected to a suitable power supply. The power supply generates an electrical potential between the two parts. As the electrode approaches the workpiece, dielectric breakdown occurs in the fluid, forming a plasma channel, and a small spark jumps.

These sparks usually strike one at a time, because it is very unlikely that different locations in the inter-electrode space have the identical local electrical characteristics which would enable a spark to occur simultaneously in all such locations. These sparks happen in huge numbers at seemingly random locations between the electrode and the workpiece. As the base metal is eroded, and the spark gap subsequently increased, the electrode is lowered automatically by the machine so that the process can continue uninterrupted. Several hundred thousand sparks occur per second, with the actual duty cycle carefully controlled by the setup parameters. These controlling cycles are sometimes known as "on time" and "off time", which are more formally defined in the literature.

The on time setting determines the length or duration of the spark. Hence, a longer on time produces a deeper cavity for that spark and all subsequent sparks for that cycle, creating a rougher finish on the workpiece. The reverse is true for a shorter on time. Off time is the period of time between sparks. A longer off time, for example, allows the flushing of dielectric fluid through a nozzle to clean out the eroded debris, thereby avoiding a short circuit. These settings can be maintained in microseconds. The typical part geometry is a complex 3D shape, often with small or odd shaped angles. Vertical, orbital, vectorial, directional, helical, conical, rotational, spin and indexing machining cycles are also used.

Wire EDM

In *wire electrical discharge machining* (WEDM), also known as *wire-cut EDM* and *wire cutting*, a thin single-strand metal wire, usually brass, is fed through the workpiece, submerged in a tank of dielectric fluid, typically deionized water. Wire-cut EDM is typically used to cut plates as thick as 300mm and to make punches, tools, and dies from hard metals that are difficult to machine with other methods. The wire, which is constantly fed from a spool, is held between upper and lower diamond guides. The guides, usually CNC-controlled, move in the *x–y* plane. On most machines, the upper guide can also move independently in the *z–u–v* axis, giving rise to the ability to cut tapered and transitioning shapes (circle on the bottom, square at the top for example). The upper guide can control axis movements in *x–y–u–v–i–j–k–l–*. This allows the wire-cut EDM to be programmed to cut very intricate and delicate shapes. The upper and lower diamond guides are usually accurate to 0.004 mm (0.16 mils), and can have a cutting path or *kerf* as small as 0.021 mm (0.83 mils) using Ø 0.02 mm (0.79 mils) wire, though the average cutting kerf that achieves the best economic cost and machining time is 0.335 mm (13.2 mils) using Ø 0.25 mm (9.8 mils) brass wire. The reason that the cutting width is greater than the width of the wire is because sparking occurs from the sides of the wire to the work piece, causing erosion. This "overcut" is necessary, for many applications it is adequately predictable and therefore can be compensated for (for instance in micro-EDM this is not often the case).

CNC Wire-cut EDM machine.

Spools of wire are long — an 8 kg spool of 0.25 mm wire is just over 19 kilometers in length. Wire diameter can be as small as 20 μm (0.79 mils) and the geometry precision is not far from ± 1 μm (0.039 mils). The wire-cut process uses water as its dielectric fluid, controlling its resistivity and other electrical properties with filters and de-ionizer units. The water flushes the cut debris away from the cutting zone. Flushing is an important factor in determining the maximum feed rate for a given material thickness. Along with tighter tolerances, multi axis EDM wire-cutting machining centers have added features such as multi heads for cutting two parts at the same time, controls for preventing wire breakage, automatic self-threading features in case of wire breakage, and programmable machining strategies to optimize the operation. Wire-cutting EDM is commonly used when low residual stresses are desired, because it does not require high cutting forces for removal of material. If the energy/power per pulse is relatively low (as in finishing operations), little change in the mechanical properties of a material is expected due to these low residual stresses, although material that hasn›t been stress-relieved can distort in the machining process. The work piece may undergo a significant thermal cycle, its severity depending on the technological parameters used. Such thermal cycles may cause formation of a recast layer on the part and residual tensile stresses on the work piece. If machining takes place after heat treatment, dimensional accuracy will not be affected by heat treat distortion.

Applications

Prototype Production

The EDM process is most widely used by the mold-making tool and die industries, but is becoming a common method of making prototype and production parts, especially in the aerospace, automobile and electronics industries in which production quantities are relatively low. In sinker EDM, a graphite, copper tungsten or pure copper electrode is machined into the desired (negative) shape and fed into the workpiece on the end of a vertical ram.

Coinage Die Making

For the creation of dies for producing jewelry and badges, or blanking and piercing (through use of a pancake die) by the coinage (stamping) process, the positive master may be made from sterling silver, since (with appropriate machine settings) the master is significantly eroded and is used only once. The resultant negative die is then hardened and used in a drop hammer to produce stamped flats from cutout sheet blanks of bronze, silver, or low proof gold alloy. For badges these flats may be further shaped to a curved surface by another die. This type of EDM is usually performed submerged in an oil-based dielectric. The finished object may be further refined by hard (glass) or soft (paint) enameling and/or electroplated with pure gold or nickel. Softer materials such as silver may be hand engraved as a refinement.

Master at top, badge die workpiece at bottom, oil jets at left (oil has been drained). Initial flat stamping will be "dapped" to give a curved surface.

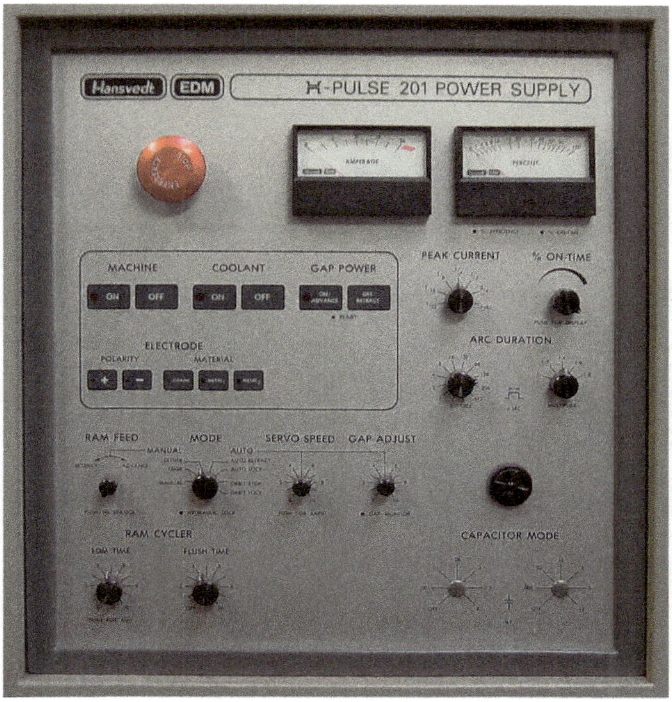

EDM control panel (Hansvedt machine). Machine may be adjusted for a refined surface (electropolish) at end of process.

Small Hole Drilling

Small hole drilling EDM is used in a variety of applications.

On wire-cut EDM machines, small hole drilling EDM is used to make a through hole in a workpiece in through which to thread the wire for the wire-cut EDM operation. A separate EDM head specifically for small hole drilling is mounted on a wire-cut machine and allows large hardened plates to have finished parts eroded from them as needed and without pre-drilling.

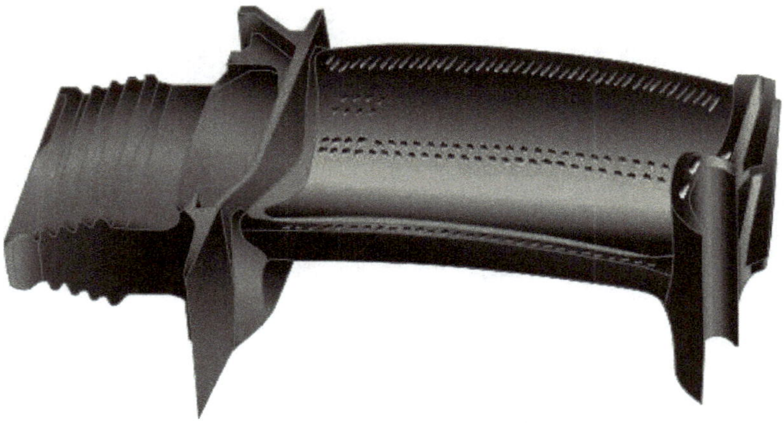

A turbine blade with internal cooling as applied in the high-pressure turbine.

Small hole drilling EDM machines.

Small hole EDM is used to drill rows of holes into the leading and trailing edges of turbine blades used in jet engines. Gas flow through these small holes allows the engines to use higher temperatures than otherwise possible. The high-temperature, very hard, single crystal alloys employed in these blades makes conventional machining of these holes with high aspect ratio extremely difficult, if not impossible.

Small hole EDM is also used to create microscopic orifices for fuel system components, spinnerets for synthetic fibers such as rayon, and other applications.

There are also stand-alone small hole drilling EDM machines with an x–y axis also known as a super drill or *hole popper* that can machine blind or through holes. EDM drills bore holes with a long brass or copper tube electrode that rotates in a chuck with a constant flow of distilled or deionized water flowing through the electrode as a flushing agent and dielectric. The electrode tubes operate like the wire in wire-cut EDM machines, having a spark gap and wear rate. Some small-hole drilling EDMs are able to drill through 100 mm of soft or through hardened steel in less than 10 seconds, averaging 50% to 80% wear rate. Holes of 0.3 mm to 6.1 mm can be achieved in this drilling operation. Brass electrodes are easier to machine but are not recommended for wire-cut operations due to eroded brass particles causing "brass on brass" wire breakage, therefore copper is recommended.

Metal Disintegration Machining

Several manufacturers produce MDM machines for the specific purpose of removing broken tools (drill bits, taps, bolts and studs) from work pieces. In this application, the process is termed "metal disintegration machining" or MDM. The metal disintegration process removes only the center of the tap, bolt or stud leaving the hole intact and allowing a part to be reclaimed.

Closed Loop Manufacturing

Closed loop manufacturing can improve the accuracy and reduce the tool costs

Advantages and Disadvantages

Advantages of EDM include machining of:

- Complex shapes that would otherwise be difficult to produce with conventional cutting tools.
- Extremely hard material to very close tolerances.
- Very small work pieces where conventional cutting tools may damage the part from excess cutting tool pressure.
- There is no direct contact between tool and work piece. Therefore, delicate sections and weak materials can be machined without perceivable distortion.
- A good surface finish can be obtained; a very good surface may be obtained by redundant finishing paths.
- Very fine holes can be attained.
- Tapered holes may be produced.

- Pipe or container internal contours and internal corners down to R .001".

Disadvantages of EDM include:

- The slow rate of material removal.
- Potential fire hazard associated with use of combustible oil based dielectrics.
- The additional time and cost used for creating electrodes for ram/sinker EDM.
- Reproducing sharp corners on the workpiece is difficult due to electrode wear.
- Specific power consumption is very high.
- Power consumption is high.
- "Overcut" is formed.
- Excessive tool wear occurs during machining.
- Electrically non-conductive materials can be machined only with specific set-up of the process.

REFERENCES

1. http://hyperphysics.phy-astr.gsu.edu/hbase/chemical/electrochem.html
2. http://www3.interscience.wiley.com:8100/legacy/college/brady/0471215171/int_dialogue/data/task_cell_notation.html
3. http://www.sparknotes.com/chemistry/electrochemistry/galvanic/section2.rhtml
4. Bard, A. J. and Faulkner, L. R. "Electrochemical methods. Fundamentals and applications", John Whiley & Sons, Inc, 2nd edition, USA, 2001
5. Todd, H. Robert; Allen, K. Dell; Alting, Leo (1994), Manufacturing Processes Reference Guide (1st ed.), Industrial Press Inc., pp. 198–199, ISBN 0-8311-3049-0.
6. Valenti, Michael, "Making the Cut." Mechanical Engineering, American Society of Mechanical Engineers, 2001. http://www.memagazine.org/backissues/membersonly/nov01/features/makcut/makcut.html accessed 2/23/2010
7. Szolc T.,; Konowrocki R.,; Michajłow M.,; Pregowska A., (2014). "An investigation of the dynamic electromechanical coupling effects in machine drive systems driven by asynchronous motors". Mechanical Systems and Signal Processing, Vol.49, pp.118-134. ISSN 0888-3270.
8. Konowrocki R.,; Szolc T.,; Pochanke A.,; Pregowska A., (2016). "An influence of the stepping motor control and friction models on precise positioning of the complex mechanical system". Mechanical Systems and Signal Processing, Vol.70-71, pp.397-413. ISSN 0888-3270.
9. Jameson 2001, p. 1.
10. Krar, Stephen F. and Gill, Arthur R. (2003) Exploring Advanced Manufacturing Technologies, Industrial Press Inc. ISBN 0831131500. p. 6.2.1

11. Jameson 2001, p. 8.

12. Jameson 2001, pp. 10–12.

13. Dulebohn, "Tracer controlled machining by electrically induced erosion", U.S. Patent 3,614,372, filed 4 December 1969, issued 19 October 1971.

14. Jameson 2001, pp. 12–17.

15. Mohri, N.; Fukuzawa, Y.; Tani, T.; Saito, N.; Furutani, K. (1996). "Assisting Electrode Method for Machining Insulating Ceramics". CIRP Annals - Manufacturing Technology. 45: 201. doi:10.1016/S0007-8506(07)63047-9.

16. Liu, Y. H.; Li, X. P.; Ji, R. J.; Yu, L. L.; Zhang, H. F.; Li, Q. Y. (2008). "Effect of technological parameter on the process performance for electric discharge milling of insulating Al2O3 ceramic". Journal of Materials Processing Technology. 208: 245. doi:10.1016/j.jmatprotec.2007.12.143.

17. Morgan, C. J.; Vallance, R. R.; Marsh, E. R. (2004). "Micro machining glass with polycrystalline diamond tools shaped by micro electro discharge machining". Journal of Micromechanics and Microengineering. 14 (12): 1687. doi:10.1088/0960-1317/14/12/013.

18. McCarthy, Willard J. and McGeough, Joseph A. "Machine tool". Encyclopædia Britannica

19. Descoeudres, Antoine (2006). Characterization of electrical discharge machining plasmas. Thèse EPFL, no 3542.

20. Weng, F. T.; Shyu, R. F.; Hsu, C. S. (2003). "Fabrication of micro-electrodes by multi-EDM grinding process". Journal of Materials Processing Technology. 140: 332. doi:10.1016/S0924-0136(03)00748-9.

21. Narasimhan, J.; Yu, Z.; Rajurkar, K. P. (2005). "Tool Wear Compensation and Path Generation in Micro and Macro EDM". Journal of Manufacturing Processes. 7: 75. doi:10.1016/S1526-6125(05)70084-0.

22. Koelsch, James (October 2009). "EDM: A Changing Competitive Calculus," Manufacturing Engineering, Society of Manufacturing Engineers

23. Han, F.; Chen, L.; Yu, D.; Zhou, X. (2006). "Basic study on pulse generator for micro-EDM". The International Journal of Advanced Manufacturing Technology. 33 (5–6): 474. doi:10.1007/s00170-006-0483-9.

24. Ferri, C.; Ivanov, A.; Petrelli, A. (2008). "Electrical measurements in μ-EDM". Journal of Micromechanics and Microengineering. 18 (8): 085007. doi:10.1088/0960-1317/18/8/085007.

25. Van Dijck, Frans (1973). Physico-mathematical analysis of the electro discharge machining process. PhD Thesis Katholieke Universiteit Leuven.

26. Dibitonto, D. D.; Eubank, P. T.; Patel, M. R.; Barrufet, M. A. (1989). "Theoretical models of the electrical discharge machining process. I. A simple cathode erosion model". Journal of Applied Physics. 66 (9): 4095. doi:10.1063/1.343994.

27. Patel, M. R.; Barrufet, M. A.; Eubank, P. T.; Dibitonto, D. D. (1989). "Theoretical models of the electrical discharge machining process. II. The anode erosion model". Journal of Applied Physics. 66 (9): 4104. doi:10.1063/1.343995.

28. Eubank, P. T.; Patel, M. R.; Barrufet, M. A.; Bozkurt, B. (1993). "Theoretical models of the electrical discharge machining process. III. The variable mass, cylindrical plasma model". Journal of Applied Physics. 73 (11): 7900. doi:10.1063/1.353942.
29. Singh, A.; Ghosh, A. (1999). "A thermo-electric model of material removal during electric discharge machining". International Journal of Machine Tools and Manufacture. 39 (4): 669. doi:10.1016/S0890-6955(98)00047-9.
30. Bilstein, Roger E. (1999). Stages to Saturn: A Technological History of the Apollo/Saturn Launch Vehicle (NASA-SP4206). DIANE Publishing. p. 145.
31. Jameson 2001.
32. Semon, G. (1975). A Practical Guide to Electro-Discharge Machining, 2nd ed. Ateliers des Charmilles, Geneva.
33. Todd, Robert H.; Allen, Dell K.; Alting, Leo (1994). Manufacturing Processes Reference Guide. Industrial Press Inc. pp. 175–179. ISBN 0-8311-3049-0.
34. ELECTRICAL DISCHARGE MACHINING (EDM). header.com
35. Kucukturk, G.; Cogun, C. (2010). "A New Method for Machining of Electrically Nonconductive Workpieces Using Electric Discharge Machining Technique". Machining Science and Technology. 14 (2): 189. doi:10.1080/10910344.2010.500497. (2010).

CHAPTER 9

Electrohydrogenesis

Chapter Contents

- **Introduction**
- **Electrohydrogenesis**
- **Photoelectrolysis**
- **Electrofuels**
- **Magnetoelectrochemistry**
- **Electrofiltration**
- **Electrodeionization**

INTRODUCTION

Electromethanogenesis is a form of electrofuel production where methane is produced by direct biological conversion from electrical current and carbon dioxide. The reduction process is carried out in a microbial electrolysis cell. A 2009 article by Cheng and Logan reports that a current capture efficiency of 96% can be achieved using a 1.0 V current.

ELECTROHYDROGENESIS

Electrohydrogenesis or biocatalyzed electrolysis is the name given to a process for generating hydrogen gas from organic matter being decomposed by bacteria. This process uses a modified fuel cell to contain the organic matter and water. A small amount, 0.2 - 0.8 V of electricity is used, the original article reports an overall energy efficiency of 288% can be achieved (this is computed relative to the amount of electricity used, waste heat lowers the overall efficiency). This work was reported by Cheng and Logan

PHOTOELECTROLYSIS

Photoelectrolysis, also known as water splitting, occurs in a photoelectrochemical cell when light is used as the energy source for the electrolysis of water, producing dihydrogen which can be used as a fuel. This process is an essential requirement of the "hydrogen economy", in which hydrogen is produced efficiently and inexpensively from natural sources without using fossil fuels. In contrast, steam reforming usually or always uses a fossil fuel to obtain hydrogen. Photoelectrolysis is sometimes known colloquially as the *hydrogen holy grail* for its potential to yield a viable alternative to petroleum as a source of energy; such an energy source would supposedly come without the sociopolitically undesirable effects of extracting and using petroleum.

Some researchers have practiced photoelectrolysis by means of a nanoscale process. Nanoscale photoelectrolysis of water could someday reach greater efficiency than that of "traditional" photoelectrolysis. Semiconductors with bandgaps smaller than 1.7 electronvolts would ostensibly be required for efficient nanoscale photoelectrolysis using light from the Sun.

Devices bases on hydrogenase have also been investigated.

Photoelectrochemistry

Photoelectrochemistry is a subfield of study within physical chemistry concerned with the interaction of light with electrochemical systems. It is an active domain of investigation. One of the pioneers of this field of electrochemistry was the German electrochemist Heinz Gerischer. The interest in this domain is high in the context of development of renewable energy conversion and storage technology.

Historical Approach

Photoelectrochemistry has been intensively studied in the 70-80s because of the first peak oil crisis. Because fossil fuels are non-renewable, it is necessary to develop processes to obtain renewable ressources and use clean energy. Artificial photosynthesis, photoelectrochemical water splitting and regenerative solar cells are of special interest in this context.

H. Gerischer, H. Tributsch, AJ. Nozik, AJ. Bard, A. Fujishima, K. Honda, PE. Laibinis, K. Rajeshwar, TJ Meyer, PV. Kamat, N.S. Lewis, R. Memming, JOM. Bockris are researchers which have contributed a lot to the field of photoelectrochemistry.

Semiconductor's Electrochemistry

Semiconductor material has a band gap and generates a pair of electron and hole per absorbed photon if the energy of the photon is higher than the band gap of the semiconductor. This property of semiconductor materials has been successfully used to convert solar energy into electrical energy by photovoltaic devices.

In photocatalysis the electron-hole pair is immediately used to drive a redox reaction but the problem is that the electron-hole pair suffer from fast recombinations. In photoelectrocatalysis, a differential potential is applied to diminish the number of recombinations between the electrons and the holes. This allows an increase in the yield of light's conversion into chemical energy.

Semiconductor-Electrolyte Interface

When a semiconductor comes into contact with a liquid (redox species), to maintain electrostatic equilibrium, there will be a charge transfer between the semiconductor and liquid phase if formal redox potential of redox species lies inside semiconductor band gap. At thermodynamic equilibrium, the Fermi level of semiconductor and the formal redox potential of redox species are aligned at the interface between semiconductor and redox species. This introduces a downward band bending in a n-type semiconductor for n-type semiconductor/liquid junction and an upward band bending in a p-type semiconductor for a p-type semiconductor/liquid junction. This characteristic of semiconductor/liquid junctions is similar to a rectifying semiconductor/metal junction or Schottky junction. Ideally to get a good rectifying characteristics at the semiconductor/liquid interface, the formal redox potential must be close to the valence band of the semiconductor for a n-type semiconductor and close to the conduction band of the semiconductor for a p-type semiconductor. The semiconductor/liquid junction has one advantage over the rectifying semiconductor/metal junction in that the light is able to travel through to the semiconductor surface without much reflection; whereas most of the light is reflected back from the metal surface at a semiconductor/metal junction. Therefore, semiconductor/liquid junctions can also be used as photovoltaic devices similar to solid state p–n junction devices. Both n-type and p-type semiconductor/liquid junctions can be used as photovoltaic devices to convert solar energy into electrical energy and are called photoelectrochemical cells. In addition, a semiconductor/liquid junction could also be used to directly convert solar energy into chemical energy by virtue of photoelectrolysis at the semiconductor/liquid junction.

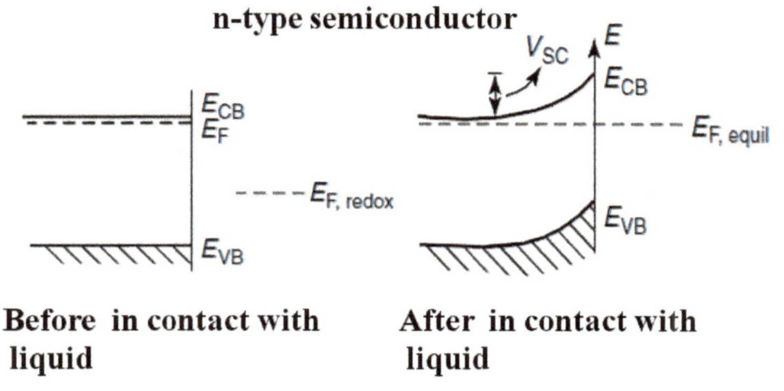

Band diagram of n-type semiconductor/liquid junction

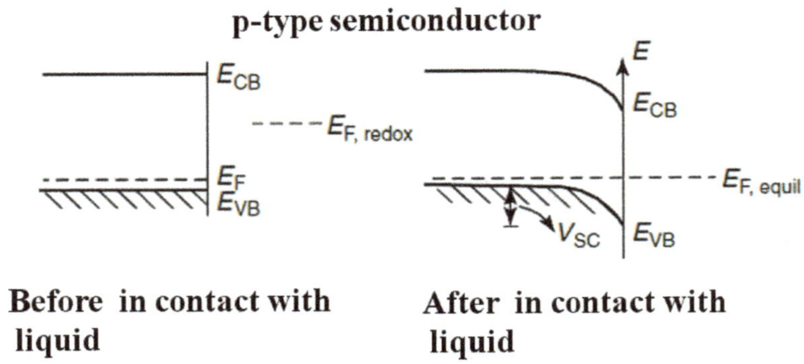

Band diagram of p-type semiconductor/liquid junction

Experimental Setup

Semiconductors are usually studied in a photoelectrochemical cell. Different configurations exist with a three electrode device. The phenomenon to study happens at the working electrode WE while the differential potential is applied between the WE and a reference electrode RE (saturated calomel, Ag/AgCl). The current is measured between the WE and the counter electrode CE (carbon vitreous, platinum gauze). The working electrode is the semiconductor material and the electrolyte is composed of a solvent, an electrolyte and a redox specie.

A UV-vis lamp is usually used to illuminate the working electrode. The photoelectrochemical cell is usually made with a quartz window because it does not absorb the light. A monochromator can be used to control the wavelength sent to the WE.

Main Absorbers Used in Photoelectrochemistry

Semiconductor IV

C(diamond), Si, Ge, SiC, SiGe

Semiconductor III-V

BN, BP, BAs, AlN, AlP, AlAs, GaN, GaP, GaAs, InN, InP, InAs...

Semiconductor II-VI

CdS, CdSe, CdTe, ZnO, ZnS, ZnSe, ZnTe, MoS_2, $MoSe_2$, $MoTe_2$, WS_2, WSe_2

Metal Oxides

TiO_2, Fe_2O_3, Cu_2O

Organic Dyes

Methylene blue...

Organometallic Dyes

Perovskites

Applications

Photoelectrochemical Splitting of Water

Photoelectrochemistry has been intensively studied in the field of hydrogen production from water and solar energy. The photoelectrochemical splitting of water was historically discovered by Fujishima and Honda in 1972 onto TiO_2 electrodes. Recently many materials have shown promising properties to split efficiently water but TiO_2 remains cheap, abundant, stable against photo-corrosion. The main problem of TiO_2 is its bandgap which is 3 or 3.2 eV according to its crystallinity (anatase or rutile). These values are too high and only the wavelength in the UV region can be absorbed. To increase the performances of this material to split water with solar wavelength, it is necessary to sensitize the TiO_2. Currently Quantum Dots sensitization is very promising but more research is needed to find new materials able to absorb the light efficiently. Recently water a splitting membrane concept has been developed. This method to split water is very similar to the principle of fuel cells but in a reverse way.

Artificial photosynthesis

Photosynthesis is The natural process that converts CO_2 using light to produce hydrocarbon compounds such as sugar. The depletion of fossil fuels encourages scientists to find alternatives to produce hydrocarbon compounds. Artificial photosynthesis is a promising method mimicking the natural photosynthesis to produce such compounds. The photoelectrochemical reduction of CO2 is much studied because of its worldwide impact. Many researchers aim to find new semiconductors to develop stable and efficient photo-anodes and photo-cathodes.

Regenerative Cells or Dye-Sensitized Solar Cell (Graetzel Cell)

Dye-sensitized solar cells or DSSCs use TiO_2 and dyes to absorb the light. This absorption induces the formation of electron-hole pairs which are used to oxidize and reduce the same redox couple, usually I^-/I_3^-. Consequently, a differential potential is created which induces a current.

Photoelectrochemical Reduction of CO_2

Photoelectrochemical reduction of CO_2 is a chemical process whereby carbon dioxide is reduced to carbon monoxide or hydrocarbons by the energy of incident light. This process needs to be catalyzed either homogeneously or heterogeneously in order to proceed, and current research is aimed at developing these catalysts, most of which are semiconducting materials. Semiconducting catalysts provide favorable electron transfer kinetics.

Motivation for research in this area is strong due to the current attention to atmospheric carbon dioxide as the reduction of carbon dioxide would be one route for removal and sequestration. Furthermore, the reduced species may prove to be a valuable feedstock for other processes. If the incident light utilized is solar in nature then this process also potentially represents energy routes which combine renewable energy with CO_2 reduction.

Background/Introduction

Semiconductor material has a band gap and generates a pair of electron and hole per absorbed photon if the energy of photon is higher than band gap of semiconductor. This property of semiconductor materials has been successfully used to convert solar energy into electrical energy by photovoltaic devices. So, semiconductor electrodes can be used for CO_2 photoelectrochemical reduction.

How Semiconductor Behaves When Come Into Contact With Liquid (Redox Species)?

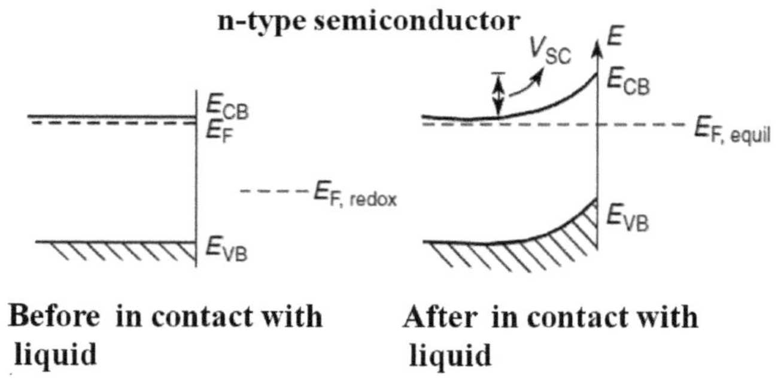

Band diagram of n-type semiconductor/liquid junction.

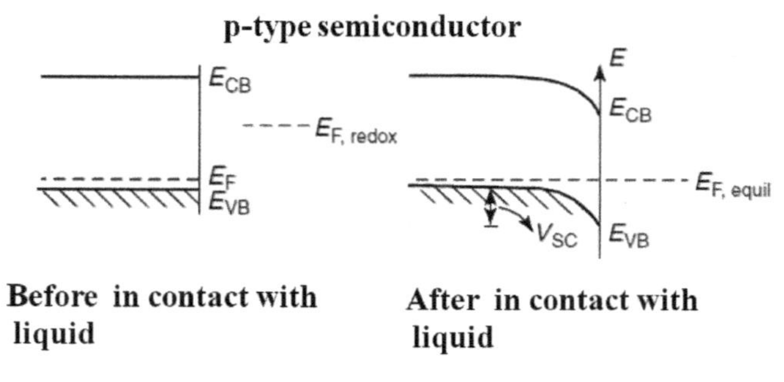

Band diagram of p-type semiconductor/liquid junction.

When a semiconductor comes into contact with a liquid (redox species), to maintain electrostatic equilibrium, there will be a charge transfer between the semiconductor and liquid phase if formal redox potential of redox species lies inside semiconductor band gap. At thermodynamic equilibrium, the Fermi level of semiconductor and the formal redox potential of redox species are aligned at the interface between semiconductor and redox species. This introduces a downward band bending in a n-type semiconductor for n-type semiconductor/liquid junction and upward band bending in a p-type semiconductor for p-type semiconductor/liquid junction. This characteristic of semiconductor/liquid junction is similar to a rectifying semiconductor/metal junction or Schottky junction. Ideally to get a good rectifying characteristics at the semiconductor/liquid interface, the formal redox potential must be close to the valence band of the semiconductor for a n-type semiconductor and close to the conduction band of the semiconductor for a p-type semiconductor. The semiconductor/liquid junction has one advantage over rectifying semiconductor/metal junction in that the light is able to travel through to semiconductor surface without much reflection; whereas most of the light is reflected back from metal surface at semiconductor/metal junction. Therefore, semiconductor/liquid junction can also be used a photovoltaic devices similar to solid state p–n junction devices. Both n-type and p-type semiconductor/liquid junction can be used as photovoltaic devices to convert solar energy into electrical energy and are called photoelectrochemical cell. In addition, a semiconductor/liquid junction could also be used to directly convert solar energy into chemical energy by virtue of photoelectrolysis at the semiconductor/liquid junction.

Why Carbon Dioxide Reduction is Important?

The catalytic conversion of CO_2 to liquid fuels is a critical goal that would positively impact the global carbon balance by recycling CO_2 into usable fuels. The challenges presented here are great, but the potential rewards are enormous. CO_2 is extremely stable molecule generally produced by fossil fuel combustion and respiration. Returning CO_2 to a useful state by activation/reduction is a scientifically challenging problem, requiring appropriate catalysts and energy input. This poses several fundamental challenges in chemical catalysis, electrochemistry, photochemistry, and semiconductor physics and engineering.

Principle

Thermodynamic Challenges

Thermodynamic potentials for the reduction of CO_2 to various products is given in the following table versus NHE at pH = 7. Single electron reduction of CO_2 to $CO_2^{\bullet-}$ radical occurs at E° = −1.90 V versus NHE at pH = 7 in an aqueous solution at 25 °C under 1 atm gas pressure. The reason behind the high negative thermodynamically unfavorable single electron reduction potential of CO_2 is the large reorganization energy between the linear molecule and bent radical anion. Proton-coupled multi-electron steps for CO_2 reductions are generally more favorable than single electron reductions, as thermodynamically more stable molecules are produced.

Summary of thermodynamic potentials of CO_2 reduction to various products	
$CO_2 + 2H^+ + 2e^- \rightarrow CO + H_2O$	$E^0 = -0.53$ V
$CO_2 + 2H^+ + 2e^- \rightarrow HCO_2H$	$E^0 = -0.61$ V
$CO_2 + 4H^+ + 4e^- \rightarrow HCHO + H_2O$	$E^0 = -0.48$ V
$CO_2 + 6H^+ + 6e^- \rightarrow CH_3OH + H_2O$	$E^0 = -0.38$ V
$CO_2 + 8H^+ + 8e^- \rightarrow CH_4 + 2H_2O$	$E^0 = -0.24$ V
$CO_2 + e^- \rightarrow CO^{-2}$	$E^0 = -1.90$ V

Kinetic Challenges

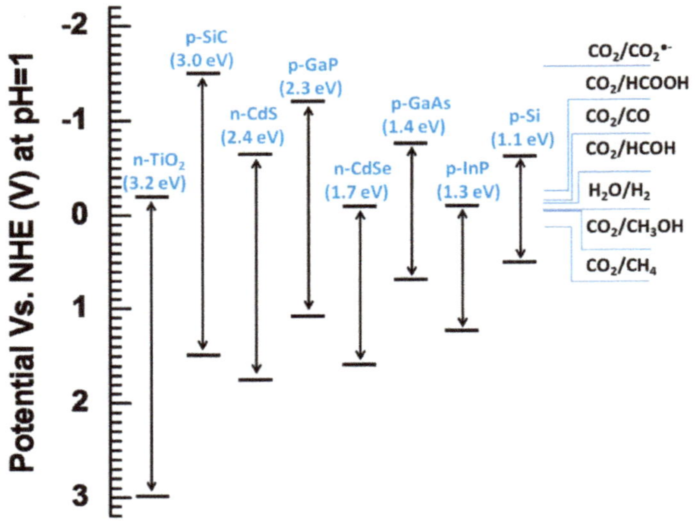

Position of conduction and valence band of several semiconductors at pH = 1 shown vs NHE. Thermodynamic potentials for CO_2 reduction to different products at pH = 1 vs NHE is shown beside the band edge positions of semiconductors.

Thermodynamically, proton coupled multiple-electron reduction of CO_2 is easier than single electron reduction. But to manage multiple proton coupled multiple-electron processes is a huge challenge kinetically. This leads to a high overpotential for electrochemical heterogeneous reduction of CO_2 to hydrocarbons and alcohols. Even further heterogeneous reduction of singly reduced $CO_2^{\bullet-}$ radical anion is difficult because of repulsive interaction between negatively biased electrode and negatively charged anion.

Is it feasible to photoelectrochemically reduce CO_2 on semiconductor surface?

Figure 1(b) shows that in case of a p-type semiconductor/liquid junction photo generated electrons are available at the semiconductor/liquid interface under illumination. The reduction of redox species happens at less negative potential on illuminated p-type semiconductor compared to metal electrode due to the band bending at semiconductor/ liquid interface. Figure 3 shows that thermodynamically, some of the proton-coupled multi-electron CO_2 reductions are within semiconductors band gap. This makes it feasible to photo-reduce CO_2 on p-type semiconductors. Various p-type semiconductors have been successfully employed for CO_2 photo reduction including p-GaP, p-CdTe, p-Si, p-GaAs, p-InP, and p-SiC. Kinetically, however, these reactions are extremely slow on given semiconductor surfaces; this leads to significant overpotential for CO_2 reduction on these semiconductor surfaces. Apart from high overpotential; these systems have a few advantages including sustainability (nothing is consumed in this system apart from light energy), direct conversion of solar energy to chemical energy, utilization of renewable energy resource for energy intensive process, stability of the process (semiconductors are really stable under illumination) etc. A different approach for photo-reduction of CO_2 involves molecular catalysts, photosensitizers and sacrificial electron donors. In this process sacrificial electron donors are consumed during the process and photosensitizers degrade under long exposure to illumination.

Solvent Effect

The photo-reduction of CO_2 on p-type semiconductor photo-electrodes has been achieved in both aqueous and non-aqueous media. Main difference between aqueous and non-aqueous media is the solubility of CO_2. The solubility of CO_2 in aqueous media at 1 atm. of CO_2 is around $\approx$ 35 mM; whereas solubility of CO_2 in methanol is around 210 mM and in acetonitrile is around 210 mM.

Aqueous Media

Halmann had first shown CO_2 photoreduction to formic acid on p-GaP as photocathode in aqueous media in 1978. Apart from several other reports of CO_2 photoreduction on p-GaP, there are other p-type semiconductors like p-GaAs, p-InP, p-CdTe, and p^+/p-Si have been successfully used for photoreduction of CO_2. The lowest potential for CO_2 photoreduction was observed on p-GaP. This may be due to high photovoltage excepted from higher band gap p-GaP (2.2 eV) photocathode. Apart from formic acid, other products observed for CO_2 photoreduction are formaldehyde, methanol and carbon monoxide. On p-GaP, p-GaAs and p^+/p-Si photocathode, the main product is formic acid with small amount of formaldehyde and methanol. However, for p-InP and p-CdTe photocathode, both carbon monoxide and formic acid are observed in similar quantities. Mechanism proposed by Hori based on CO_2 reduction on metal electrodes predicts formation of both formic acid (in case of no adsorption of singly reduced $CO_2^{\bullet-}$ radical anion to the surface) and carbon monoxide (in case of adsorption of singly reduced $CO_2^{\bullet-}$ radical anion to the surface) in aqueous media. This same mechanism can be evoked to explain the formation of mainly formic acid on p-GaP, p-GaAs and p^+/p-Si photocathode owing to no adsorption of singly reduced $CO_2^{\bullet-}$ radical anion to the surface. In case of p-InP and p-CdTe photocathode, partial adsorption of $CO_2^{\bullet-}$ radical anion leads to formation of both carbon monoxide and formic acid. Low catalytic current density for CO_2 photoreduction and competitive hydrogen generation are two major drawbacks of this system.

Non-Aqueous Media

Maximum catalytic current density for CO_2 reduction that can be achieved in aqueous media is only 10 mA cm^{-2} based solubility of CO_2 and diffusion limitations. The integrated maximum photocurrent under Air Mass 1.5 illumination, in the conventional Shockley-Quiesser limit for solar energy conversion for p-Si (1.12 eV), p-InP (1.3 eV), p-GaAs (1.4 eV), and p-GaP (2.3 eV) are 44.0 mA cm^{-2}, 37.0 mA cm^{-2}, 32.5 mA cm^{-2} and 9.0 mA cm^{-2}, respectively. Therefore, non-aqueous media such as DMF, acetonitrile, methanol are explored as solvent for CO_2 electrochemical reduction. In addition, Methanol has been industrially used as a physical absorber of CO_2 in the Rectisol method. Similarly to aqueous media system, p-Si, p-InP, p-GaAs, p-GaP and p-CdTe are explored for CO_2 photoelectrochemical reduction. Among these, p-GaP has lowest overpotential, whereas, p-CdTe has moderate overpotential but high catalytic current density in DMF with 5% water mixture system. Main product of CO_2 reduction in non-aqueous media is carbon monoxide. Competitive hydrogen generation is minimized in non-aqueous media. Proposed mechanism for CO_2 reduction to CO in non-aqueous media involves single electron reduction of CO_2 to $CO_2^{\bullet-}$ radical anion and adsorption of radical anion to surface followed by disproportionate reaction between unreduced CO_2 and $CO_2^{\bullet-}$ radical anion to form CO_3^{2-} and CO.

ELECTROFUELS

Electrofuels are an emerging class of carbon-neutral drop-in replacement fuels that are made by storing electrical energy from renewable sources in the chemical bonds of liquid or gas fuels. The primary targets are butanol, biodiesel, and hydrogen, but include other alcohols and carbon-containing gasses such as methane and butane.

A primary source of funding for research on liquid electrofuels for transportation is the Electrofuels Program of the Advanced Research Projects Agency-Energy (ARPA-E), headed by Eric Toone. ARPA-E, created in 2009 under President Obama's Secretary of Energy Steven Chu, is the Department of Energy's (DOE) attempt to duplicate the effectiveness of the Defense Advanced Research Projects Agency, DARPA. Examples of projects funded under this program include OPX Biotechnologies' biodiesel effort led by Michael Lynch and Derek Lovley's work on microbial electrosynthesis at the University of Massachusetts Amherst, which reportedly produced the first liquid electrofuel using CO2 as the feedstock. Descriptions of all ARPA-E Electrofuels Program research projects can be found at the ARPA-E Electrofuels Program website.

The first Electrofuels Conference, sponsored by the American Institute of Chemical Engineers was held in Providence, RI in November 2011. At that conference, Director Eric Toone stated that "Eighteen months into the program, we know it works. We need to know if we can make it matter." Several groups are beyond proof-of-principle, and are working to scale up cost-effectively.

Electrofuels have the potential to be disruptive if carbon-neutral electrofuels can be cheaper than petroleum fuels, and chemical feedstocks produced by electrosynthesis cheaper than those refined from crude oil. Electrofuels also have a great potential to alter the renewable energy landscape, as electrofuels allow renewables from all sources to be stored conveniently as a liquid fuel.

As of 2014, prompted by the fracking boom, ARPA-E's focus has moved from electrical feedstocks to natural-gas based feedstocks, and thus away from electrofuels.

MAGNETOELECTROCHEMISTRY

Magnetoelectrochemistry is a branch of electrochemistry dealing with magnetic effects in electrochemistry.

History

These effects have been supposed to exist since the time of Michael Faraday. There have also been observations on the existence of Hall effect in electrolytes. Until these observations, magnetoelectrochemistry was an esoteric curiosity, though this field has had a rapid development in the past years and is now an active area of research. Other scientific fields which contributed to the development of magnetoelectrochemistry are magnetohydrodynamics and convective diffusion theory.

Effects of Magnetic Field

There are three types of magnetic effects in electrochemistry:

- on electrolytes
- on mass transfer
- on metal deposition

ELECTROFILTRATION

Electrofiltration is a method that combines membrane filtration and electrophoresis in a dead-end process.

Electrofiltration is regarded as an appropriate technique for concentration and fractionation of biopolymers. The film formation on the filter membrane which hinders filtration can be minimized or completely avoided by the application of electric field, improving filtration's performance and increasing selectivity in case of fractionation. This approach reduces significantly the expenses for downstream processing in bioprocesses.

Technique

Electrofiltration is a technique for separation and concentration of colloidal substances - for instance biopolymers. The principle of electrofiltration is based on overlaying electric field on a standard dead-end filtration. Thus the created polarity facilitates electrophoretic force which is opposite to the resistance force of the filtrate flow and directs the charged biopolymers. This provides extreme decrease in the film formation on the micro- or ultra-filtration membranes and the reduction of filtration time from several hours by standard filtration to a few minutes by electrofiltration. In comparison to cross-flow filtration electrofiltration exhibits not only increased permeate flow but also guarantees reduced shear force stress which qualifies it as particularly mild technique for separation of biopolymers that are usually unstable.

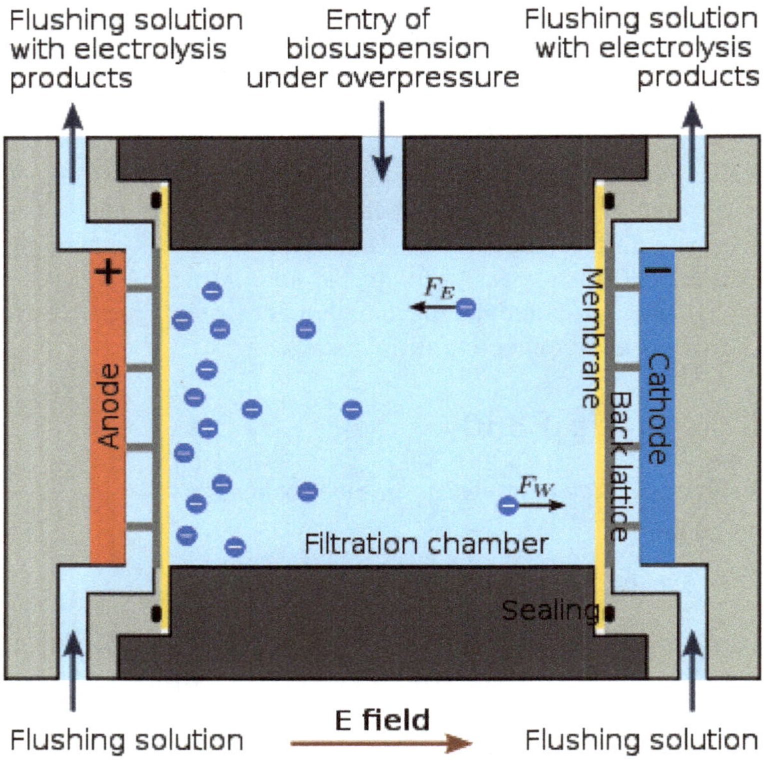

Schematic representation of an electrofiltration chamber.

The promising application in purification of biotechnological products is based on the fact that biopolymers are difficult for filtration but on the other hand they are usually charged as a result of the presence of amino and carboxyl groups. The objective of electrofiltration is to prevent the formation of filter cake and to improve the filtration kinetic of products difficult to filtrate.

The electrophoresis of the particles and the electro-osmosis become essential when the filtration process is overlaid with electric field. By electrofiltration the conventional filtration is overlaid with an electric field (DC) which works parallel with the filtrate's flow direction. When the electrophoretic force F_E, oppositely directed to flow, overruns the hydrodynamic resistance force F_W, the charged particles migrate from the filter medium, thus reducing significantly the thickness of the filter cake on the membrane.

When the solid particles, subject to separation, are negatively charged they migrate towards the anode (positive pole) and deposit on the filter cloth situated there. As a result, on the cathode side's membrane (negative pole) there is only a very thin film allowing nearly the whole filtrate to efflux through this membrane.

Figure 1 presents schematic description of electrofiltration chamber with flushing electrodes. For the flushing circulation a buffer solution is used. This approach has been patented.

Fundamental

Filter cake of xanthan on filter plate.

The hydrodynamic resistance force is evaluated following the Stokes' law.

$$F_W = 6 \cdot \pi \cdot \eta \cdot r_H \cdot \nu$$

The electrophoretic force is evaluated following the Coulomb's law.

$$F_E = 4 \cdot \pi \cdot \varepsilon_0 \cdot \varepsilon_r \cdot r_H \cdot \zeta \cdot E$$

In these equations r_H presents the hydrodynamic radius of the colloids, ν – the speed of electrophoretic migration, η – the dynamic viscosity of the solutions, ε_0 – dielectric constant in vacuum, ε_r is water's relative dielectric constant at 298 K, ζ is the zeta potential, E is the electric field. The hydrodynamic radius is the sum of particles' radiuses and the stationary solvent interface.

By steady state electrophoretic migration of charged colloids the electrophoretic force and the hydrodynamic resistance force are in equilibrium, described by:

$$F_W + F_E = 0$$

Those effects influence the electrofiltration of biopolymers, which could be also charged, not only by the hydrodynamic resistance force but also by the electric field force. Focusing on the cathode side reveals that the negatively charged particles are affected by the electric field force, which is opposite to the hydrodynamic resistance force. In this manner the formation of filter cake on this side is impeded or in ideal situation filter cake is not formed at all. In this case the electric field is referred as critical electric field E_{crit}. As a result of the equilibrium of those forces, liquids subjected to the influence of electric force become charged. In addition to the applied hydraulic pressure ΔpH the process is influenced also by the electro-osmotic pressure P_e.

Modifying the Darcy's basic equation, describing filter cake formation, with electro-kinetic effects by integration under assumption of using the constants of electro-osmotic pressure P_e, the critical electric field E_{krit} and the electric field E results:

$$\frac{t}{V_L} = \frac{\eta \cdot \alpha_c \cdot c \cdot \frac{(E_{crit} - E)}{E_{crit}}}{2 \cdot (\Delta P_H + P_e) \cdot A^2} \cdot V_L$$

In this equation α_c represents the mass specific cake resistance, c – concentration, A is filtration's surface, V_L – volume of the filtrate, ΔP_H is the hydraulic pressure.

Previous scientific works conducted in the Dept. of Bioprocess Engineering, Institute of Engineering in Life Sciences, University of Karlsruhe demonstrated that electrofiltration is effective for the concentration of charged biopolymers. Very promising results concerning purification of the charged polysaccharide xanthan are already obtained. Figure 2 represents xanthan filter cake.

ELECTRODEIONIZATION

Electrodeionization is a water treatment technology that utilizes an electricity, ion exchange membranes and resin to deionize water and separate dissolved ions (impurities) from water. It differs from other water purification technologies in that it is done without the use of chemical treatments and is usually a polishing treatment to reverse osmosis (RO). There are also EDI units that are often referred to as continuous electrodeionization (CEDI) since the electric current regenerates the resin mass continuously. CEDI technique can achieve very high purity, with conductivity below 0.1 μS/cm.

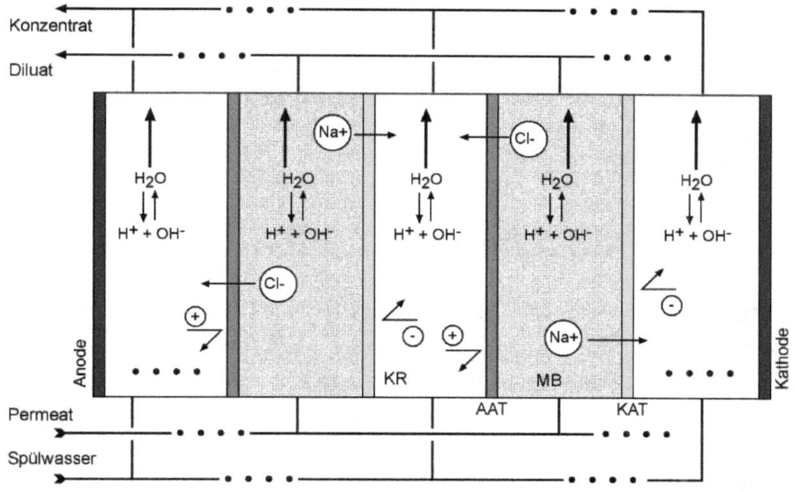

Schematic diagram of an EDI module

Applications

When fed with low TDS feed (e.g., feed purified by RO), the product can reach very high purity levels (e.g., 18 Megohms/cm). The ion exchange resins act to retain the ions, allowing these to be transported across the ion exchange membranes. The main usage of EDI technology such as that supplied by Ionpure,E-cell and SnowPure are in electronics, pharmaceutical, and power generation applications.

One important aspect in the water treatment application is that some types of EDI needs to have feed water that is free from CO2, as well as other dissolved gasses, since these are in a dissolved form they will put unnecessary strain on the EDI unit and will reduce performance.

Theory

An electrode in an electrochemical cell is referred to as either an anode or a cathode, terms that were coined by Michael Faraday. The anode is defined as the electrode at which electrons leave the cell and oxidation occurs, and the cathode as the electrode at which electrons enter the cell and reduction occurs. Each electrode may become either the anode or the cathode depending on the voltage applied to the cell. A bipolar electrode is an electrode that functions as the anode of one cell and the cathode of another cell.

Each cell consists of an electrode and an electrolyte with ions that undergo either oxidation or reduction. An electrolyte is a substance containing free ions that behaves as an electrically conductive medium. Because they generally consist of ions in solution, electrolytes are also known as ionic solutions, but molten electrolytes and solid electrolytes are also possible. They are sometimes referred to in abbreviated jargon as lytes.

Water is passed between an anode (positive electrode) and a cathode (negative electrode). Ion-selective membranes allow the positive ions to separate from the water toward the negative electrode and the negative ions toward the positive electrode. High purity deionized water results.

Electrochemical Reduction of Carbon Dioxide

The electrochemical reduction of carbon dioxide (ERC) is the conversion of carbon dioxide to more reduced chemical species using electrical energy. The first examples of electrochemical reduction of carbon dioxide are from the 19th century, when carbon dioxide was reduced to carbon monoxide using a zinc cathode. Research in this field intensified in the 1980s following the oil embargoes of the 1970s. Electrochemical reduction of carbon dioxide represents a possible means of producing chemicals or fuels, converting carbon dioxide (CO_2) to organic feedstocks such as formic acid (HCOOH), methanol (CH_3OH), ethylene (C_2H_4), methane (CH_4), and carbon monoxide (CO).

Chemicals from Carbon Dioxide

In carbon fixation plants convert carbon dioxide into sugars, from which many biosynthetic pathways originate. The catalyst responsible for this conversion, RuBisCo, is the most common protein on earth. Some anaerobic organisms employ enzymes to convert CO_2 to carbon monoxide, from which fatty acids can be made.

In industry, a few products are made from CO_2, including urea, salicylic acid, methanol, and certain inorganic and organic carbonates. In the laboratory, carbon dioxide is sometimes used to prepare carboxylic acids. No electrochemical process involving CO_2 has been commercialized.

Electrocatalysis

The electrochemical reduction of carbon dioxide to CO is usually described as:

$$CO_2 + 2\ H^+ + 2\ e^- \rightarrow CO + H_2O$$

The redox potential for this reaction is similar to that for hydrogen evolution in aqueous electrolytes, thus electrochemical reduction of CO2 is usually competitive with hydrogen evolution reaction.

Electrochemical methods have gained significant attention: 1)at ambient pressure and room temperature; 2)in connection with renewable energy sources competitive controllability, modularity and scale-up is relatively simple. The electrochemical reduction or electrocatalytic conversion of CO_2 can produce value-added chemicals such methane, ethylene, ethane, etc. and the products are mainly depends on the selected catalysts and operating potentials (applying reduction voltage).

Although an electrochemical route to CO (or other chemicals) has not been commercialized, a variety of homogeneous and heterogeneous catalysts have been evaluated. Many such processes are assumed to operate via the intermediacy of metal carbon dioxide complexes. Generally speaking, the processes developed to date either have poor thermodynamic efficiency (high overpotential), low current efficiency, low selectivity, slow kinetics, and/or poor stability.

Bioelectrochemical Reactor

Bioelectrochemical reactors are a type of bioreactor where bioelectrochemical processes can take place. They are used in bioelectrochemical syntheses, environmental remediation and electrochemical energy conversion. Examples of bioelectrochemical reactors include microbial electrolysis cells, microbial fuel cells and enzymatic biofuel cells and electrolysis cells, microbial electrosynthesis cells, and biobatteries. This bioreactor is divided in two parts: The anode, where the oxidation reaction takes place; And the cathode, where the reduction occurs.

History

In 1911 M. Potter described how microbial conversions could create reducing power,and thus electric current. Twenty years later Cohen (1931) investigate the capacity of bacteria to produce an electrical flow and he noted that the main limitation is the small capacity of microorganism to current generation. It tooks until the 60's since Berg and Canfield (1964) build the first Microbial Fuel Cell (MFC). Nowadays the investigation on bioelectrochemical reactors is increasing exponentially. These devices have real application on fields like water treatment, energy production and storage, resources production recycling and recovery.

Principles

Electron current is inherent to the microbial metabolism. Microorganism transfer electrons from an electron donor (lower potential species)to an electron acceptor (higher potential species) If the electron acceptor is an external ion or molecule, the process is called respiration. If the process is internal, electron transfer is called fermentation. The microorganism attempt to maximize their energy gain by selecting the electron acceptor with the highest potential available. In nature mainly mineral containing iron or manganese oxides are being reduced. Often soluble electron acceptors are depleted in the microbial environment. The microorganism can also maximize their energy selecting a good electron donor which can be easily metabolized. These processes are done by extracellular electron transfer (EET). The theoretical energy gain ΔG for microorganism relates directly the potential difference between the electron acceptor and the donor. But the inefficiencies like internal resistances will decrease this energy gain. The advantage of these devices is their high selectivity and in high speed processes limited by kinetic factors. The most common studied species are Shewanella oneidensis and Geobacter sulfurreducens. However, more species have been studied in recent years.

In March 25, 2013, Scientists at the University of East Anglia were able to transfer electrical charge by letting the bacteria touch on a metal or mineral surface. The research shows that it is possible to ‹tether› bacteria directly to electrodes.

Microbial Electrosynthesis

Microbial electrosynthesis is a form of microbial electrocatalysis in which electrons are supplied to living microorganisms via a cathode in an electrochemical cell by applying an electric current. The electrons are then used by the microorganisms to reduce carbon dioxide to yield industrially relevant products. The electric current would ideally be produced by a renewable source of power. This process is the opposite to that employed in a microbial fuel cell, in which microorganisms transfer electrons from the oxidation of compounds to an anode to generate an electric current.

Comparison to microbial Electrolysis Cells

Microbial electrosynthesis is related to microbial electrolysis cells (MEC). Both use the interactions of microorganisms with a cathode to reduce chemical compounds, but differ in the source of power for this process. In microbial electrosynthesis electrons are supplied solely by the external electrical power source. In MECs, an electrical power source is used to augment the electrical potential produced by the microorganisms consuming a source of chemical energy such as acetic acid. The combined potential provided by the power source and the microorganisms is then sufficient to for example reduce hydrogen ions to molecular hydrogen.

Applications

Microbial electrosynthesis may be used to produce fuel from carbon dioxide using electrical energy generated by either traditional power stations or renewable electricity generation. It may also be used to produce speciality chemicals such as drug precursors through microbially assisted electrocatalysis.

Microbial Fuel Cell

A microbial fuel cell (MFC), or biological fuel cell, is a bio-electrochemical system that drives an electric current by using bacteria and mimicking bacterial interactions found in nature. MFCs can be grouped into two general categories: mediated and unmediated. The first MFCs, demonstrated in the early 20th century, used a mediator: a chemical that transfers electrons from the bacteria in the cell to the anode. Unmediated MFCs emerged in the 1970s; in this type of MFC the bacteria typically have electrochemically active redox proteins such as cytochromes on their outer membrane that can transfer electrons directly to the anode. In the 21st century MFCs started to find a commercial use in wastewater treatment.

History

The idea of using microbes to produce electricity was conceived in the early twentieth century. M. C. Potter initiated the subject in 1911. Potter managed to generate electricity from *Escherichia coli*, but the work received little coverage. In 1931, Branet Cohen created microbial half fuel cells that, when connected in series, were capable of producing over 35 volts with only a current of 2 milliamps.

A study by DelDuca et al. used hydrogen produced by the fermentation of glucose by *Clostridium butyricum* as the reactant at the anode of a hydrogen and air fuel cell. Though the cell functioned, it was unreliable owing to the unstable nature of hydrogen production by the micro-organisms. This issue was resolved by Suzuki et al. in 1976, who produced a successful MFC design a year later.

In the late 1970s little was understood about how microbial fuel cells functioned. The idea was studied by Robin M. Allen and later by H. Peter Bennetto. People saw the fuel cell as a possible method for the generation of electricity for developing countries. Bennetto's work, starting in the early 1980s, helped build an understanding of how fuel cells operate and he was seen by many as the topic's foremost authority.

In May 2007, the University of Queensland, Australia completed a prototype MFC as a cooperative effort with Foster's Brewing. The prototype, a 10 L design, converted brewery wastewater into carbon dioxide, clean water and electricity. The group had plans to create a pilot-scale model for an upcoming international bio-energy conference.

Definition

A microbial fuel cell (MFC) is a device that converts chemical energy to electrical energy by the action of microorganisms. These electrochemical cells are constructed using either a bioanode and/or a biocathode. Most MFCs contain a membrane to separate the compartments of the anode (where oxidation takes place) and the cathode (where reduction takes place). The electrons produced during oxidation are transferred directly to an electrode or, to a redox mediator species. The electron flux is moved to the cathode. The charge balance of the system is compensated by ionic movement inside the cell, usually across an ionic membrane. Most MFCs use an organic electron donor that is oxidised to produce CO_2, protons and electrons. Other electron donors have been reported, such as sulphur compounds or hydrogen. The cathode reaction uses a variety of electron acceptors that includes the reduction of oxygen as the most studied process. However, other electron acceptors have been studied, including metal recovery by reduction, water to hydrogen, nitrate reduction and sulfate reduction.

Applications

Power Generation

MFCs are attractive for power generation applications that require only low power, but where replacing batteries may be impractical, such as wireless sensor networks.

Virtually any organic material could be used to feed the fuel cell, including coupling cells to wastewater treatment plants. MFCs are a clean and efficient method of energy production. Chemical process wastewater and synthetic wastewater have been used to produce bioelectricity in dual- and single-chamber mediatorless MFCs (uncoated graphite electrodes).

Higher power production was observed with a biofilm-covered graphite anode. Fuel cell emissions are well under regulatory limits. MFCs use energy more efficiently than standard internal combustion engines, which are limited by the Carnot Cycle. In theory, an MFC is capable of energy efficiency far beyond 50%. Rozendal obtained energy conversion to hydrogen 8 times that of conventional hydrogen production technologies.

However; MFCs can also work at a smaller scale. Electrodes in some cases need only be 7 μm thick by 2 cm long. such that an MFC can replace a battery. It provides a renewable form of energy and does not need to be recharged.

MFCs operate well in mild conditions, 20 °C to 40 °C and also at pH of around 7. They lack the stability required for long-term medical applications such as in pacemakers.

Power stations can be based on aquatic plants such as algae. If sited adjacent to an existing power system, the MFC system can share its electricity lines.

Education

Soil-based microbial fuel cells serve as educational tools, as they encompass multiple scientific disciplines (microbiology, geochemistry, electrical engineering, etc.) and can be made using commonly available materials, such as soils and items from the refrigerator. Kits for home science projects and classrooms are available.

Biosensor

The current generated from a microbial fuel cell is directly proportional to the energy content of wastewater used as the fuel. MFCs can measure the solute concentration of wastewater (i.e., as a biosensor).

Wastewater is commonly assessed for its biochemical oxygen demand (BOD) values. BOD values are determined by incubating samples for 5 days with proper source of microbes, usually activated sludge collected from wastewater plants.

An MFC-type BOD sensor can provide real-time BOD values. Oxygen and nitrate are preferred electron acceptors over the electrode, reducing current generation from an MFC. MFC BOD sensors underestimate BOD values in the presence of these electron acceptors. This can be avoided by inhibiting aerobic and nitrate respiration in the MFC using terminal oxidase inhibitors such as cyanide and azide. Such BOD sensors are commercially available.

The United States Navy is considering microbial fuel cells for environmental sensors. The use of microbial fuel cells to power environmental sensors would be able to provide power for longer periods and enable the collection and retrieval of undersea data without a wired infrastructure. The energy created by these fuel cells is enough to sustain the sensors after an initial startup time. Due to undersea conditions (high salt concentrations, fluctuating temperatures and limited nutrient supply), the Navy may deploy MFCs with a mixture of salt-tolerant microorganisms. A mixture would allow for a more complete utilization of available nutrients. *Shewanella oneidensis* is their primary candidate, but may include other heat- and cold-tolerant *Shewanella spp.*

Biorecovery

In 2010, A. ter Heijne et al. constructed a device capable of producing electricity and reducing Cu (II) (ion) to copper metal.

Microbial electrolysis cells have been demonstrated to produce hydrogen.

Wastewater Treatment

MFCs are used in water treatment to harvest energy utilizing anaerobic digestion. The process can also reduce pathogens. However, it requires temperatures upwards of 30 degrees C and requires an extra step in order to convert biogas to electricity. Spiral spacers may be used to increase electricity generation by creating a helical flow in the MFC. Scaling MFCs is a challenge because of the power output challenges of a larger surface area.

Types

Mediated

Most microbial cells are electrochemically inactive. Electron transfer from microbial cells to the electrode is facilitated by mediators such as thionine, methyl viologen, methyl blue, humic acid and neutral red. Most available mediators are expensive and toxic.

Mediator-Free

Mediator-free microbial fuel cells use electrochemically active bacteria to transfer electrons to the electrode (electrons are carried directly from the bacterial respiratory enzyme to the electrode). Among the electrochemically active bacteria are *Shewanella putrefaciens*, *Aeromonas hydrophila* and others. Some bacteria are able to transfer their electron production via the pili on their external membrane. Mediator-free MFCs are less well-characterized, such as the strain of bacteria used in the system, type of ion-exchange membrane and system conditions (temperature, pH, etc.)

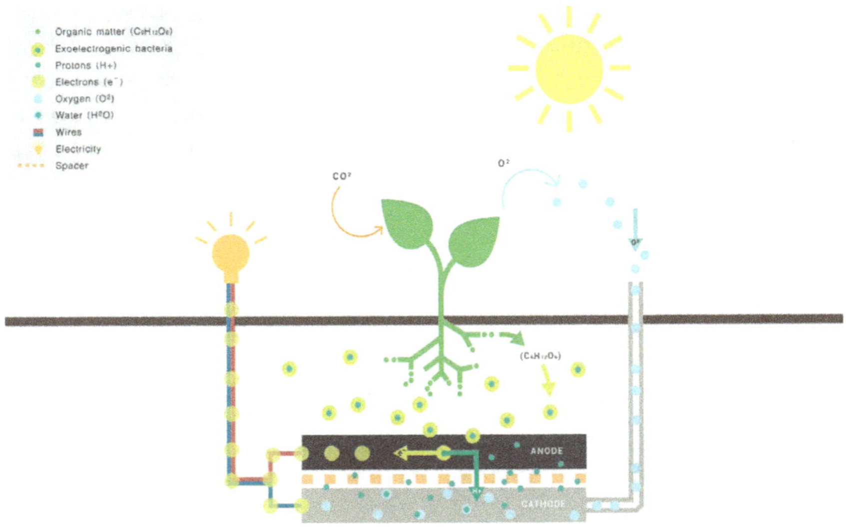

A plant microbial fuel cell (PMFC).

Mediator-free microbial fuel cells can run on wastewater and derive energy directly from certain plants. This configuration is known as a plant microbial fuel cell. Possible plants include reed sweetgrass, cordgrass, rice, tomatoes, lupines and algae. Given that the power is derived from living plants (*in situ*-energy production), this variant can provide ecological advantages.

Microbial Electrolysis

One variation of the mediator-less MFC is the microbial electrolysis cell (MEC). While MFCs produce electric current by the bacterial decomposition of organic compounds in water, MECs partially reverse the process to generate hydrogen or methane by applying a voltage to bacteria. This supplements the voltage generated by the microbial decomposition of organics, leading to the electrolysis of water or methane production. A complete reversal of the MFC principle is found in microbial electrosynthesis, in which carbon dioxide is reduced by bacteria using an external electric current to form multi-carbon organic compounds.

Soil-Based

Soil-based microbial fuel cells adhere to the basic MFC principles, whereby soil acts as the nutrient-rich anodic media, the inoculum and the proton exchange membrane (PEM). The anode is placed at a particular depth within the soil, while the cathode rests on top the soil and is exposed to air.

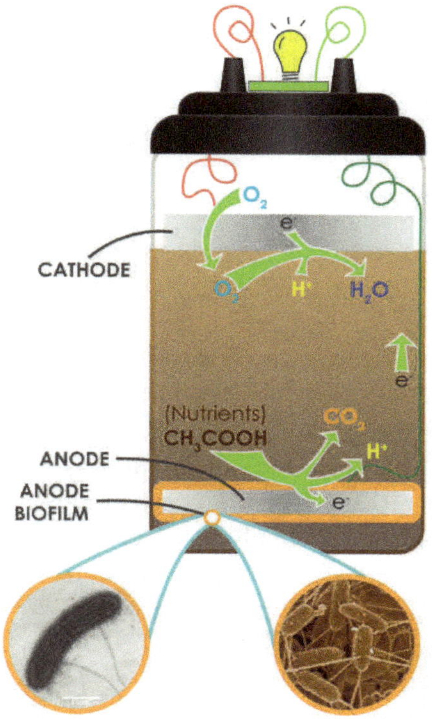

A soil-based MFC.

Soils naturally teem with diverse microbes, including electrogenic bacteria needed for MFCs, and are full of complex sugars and other nutrients that have accumulated from plant and animal material decay. Moreover, the aerobic (oxygen consuming) microbes present in the soil act as an oxygen filter, much like the expensive PEM materials used in laboratory MFC systems, which cause the redox potential of the soil to decrease with greater depth. Soil-based MFCs are becoming popular educational tools for science classrooms. Sediment microbial fuel cells (SMFCs) have been applied for wastewater treatment. Simple SMFCs can generate energy while decontaminating wastewater. Most such SMFCs contain plants to mimic constructed wetlands. By 2015 SMFC tests had reached more than 150 l.

In 2015 researchers announced an SMFC application that extracts energy and charges a battery. Salts dissociate into positively and negatively charged ions in water and move and adhere to the respective negative and positive electrodes, charging the battery and making it possible to remove the salt effecting *microbial capacitive desalination*. The microbes produce more energy than is required for the desalination process.

Phototrophic Biofilm

Phototrophic biofilm MFCs (ner) use a phototrophic biofilm anode containing photosynthetic microorganism such as chlorophytacandyanophyta.They carry out photosynthesis and thus produce organic metabolites and donate electrons.

One study found that PBMFCs display a power density sufficient for practical applications.

The sub-category of phototrophic MFCs that use purely oxygenic photosynthetic material at the anode are sometimes called biological photovoltaic systems.

Nanoporous Membrane

The United States Naval Research Laboratory developed nanoporous membrane microbial fuel cells that use a non-PEM to generate passive diffusion within the cell. The membrane is a nonporous polymer filter (nylon, cellulose, or polycarbonate). It offers comparable power densities to Nafion (a well known PEM) with greater durability. Porous membranes allow passive diffusion thereby reducing the necessary power supplied to the MFC in order to keep the PEM active and increasing the total energy output.

MFCs that do not use a membrane can deploy anaerobic bacteria in aerobic environments. However, membrane-less MFCs experience cathode contamination by the indigenous bacteria and the power-supplying microbe. The novel passive diffusion of nanoporous membranes can achieve the benefits of a membrane-less MFC without worry of cathode contamination.

Nanoporous membranes are also eleven times cheaper than Nafion (Nafion-117, $0.22/cm^2 vs. polycarbonate, <$0.02/cm^2).

Ceramic Membrane

PEM membranes can be replaced with ceramic materials. Ceramic membrane costs can be as low as $5.66 /m^2. The macroporous structure of ceramic membranes allows good transport of ionic species.

The materials that have been successfully employed in ceramic MFCs are earthenware, alumina, mullite, pyrophyllite and terracotta.

Generation Process

When microorganisms consume a substance such as sugar in aerobic conditions, they produce carbon dioxide and water. However, when oxygen is not present, they produce carbon dioxide, protons and electrons, as described below:

$$C_{12}H_{22}O_{11} + 13H_2O \rightarrow 12CO_2 + 48H^+ + 48e^- \quad (\textit{Eqt. 1})$$

Microbial fuel cells use inorganic mediators to tap into the electron transport chain of cells and channel electrons produced. The mediator crosses the outer cell lipid membranes and bacterial outer membrane; then, it begins to liberate electrons from the electron transport chain that normally would be taken up by oxygen or other intermediates.

The now-reduced mediator exits the cell laden with electrons that it transfers to an electrode; this electrode becomes the anode. The release of the electrons recycles the mediator to its original oxidised state, ready to repeat the process. This can happen only under anaerobic conditions; if oxygen is present, it will collect the electrons, as it has greater electronegativity.

In MFC operation, the anode is the terminal electron acceptor recognized by bacteria in the anodic chamber. Therefore, the microbial activity is strongly dependent on the anode's redox potential. A Michaelis-Menten curve was obtained between the anodic potential and the power output of an acetate-driven MFC. A critical anodic potential seems to provide maximum power output.

Potential mediators include natural red, methylene blue, thionine and resorufin.

Organisms capable of producing an electric current are termed exoelectrogens. In order to turn this current into usable electricity, exoelectrogens have to be accommodated in a fuel cell.

The mediator and a micro-organism such as yeast, are mixed together in a solution to which is added a substrate such as glucose. This mixture is placed in a sealed chamber to stop oxygen entering, thus forcing the micro-organism to undertake anaerobic respiration. An electrode is placed in the solution to act as the anode.

In the second chamber of the MFC is another solution and the positively charged cathode. It is the equivalent of the oxygen sink at the end of the electron transport chain, external to the biological cell. The solution is an oxidizing agent that picks up the electrons at the cathode. As with the electron chain in the yeast cell, this could be a variety of molecules such as oxygen, although a more convenient option is a solid oxidizing agent, which requires less volume.

Connecting the two electrodes is a wire (or other electrically conductive path). Completing the circuit and connecting the two chambers is a salt bridge or ion-exchange membrane. This last feature allows the protons produced, as described in *Eqt. 1,* to pass from the anode chamber to the cathode chamber.

The reduced mediator carries electrons from the cell to the electrode. Here the mediator is oxidized as it deposits the electrons. These then flow across the wire to the second electrode, which acts as an electron sink. From here they pass to an oxidizing material.

REFERENCES

1. Hydrogen production through biocatalyzed electrolysis
2. Cheng, S; Logan, BE (November 2007). "Sustainable and efficient biohydrogen production via electrohydrogenesis". Proc. Natl. Acad. Sci. U.S.A. 104: 18871–3. doi:10.1073/pnas.0706379104. PMC 2141869Freely accessible. PMID 18000052.
3. Shaoan Cheng; Defeng Xing; Douglas F. Call; Bruce E. Logan (March 26, 2009). "Direct Biological Conversion of Electrical Current into Methane by Electromethanogenesis". Environ. Sci. Technol. American Chemical Society. 43 (10): 3953–8. doi:10.1021/es803531g. PMID 19544913. Retrieved 2009-04-07.
4. Crabtree, G. W.; Dresselhaus, M. S.; Buchanan, M. V. (2004). "The Hydrogen Economy". Physics Today. 57: 39. Bibcode:2004PhT....57l..39C. doi:10.1063/1.1878333.
5. Parkin, Alison (2014). "Chapter 5. Understanding and Harnessing Hydrogenases, Biological Dihydrogen Catalysts". In Peter M.H. Kroneck and Martha E. Sosa Torres. The Metal-Driven Biogeochemistry of Gaseous Compounds

in the Environment. Metal Ions in Life Sciences. 14. Springer. pp. 99–124. doi:10.1007/978-94-017-9269-1_5.

6. Benson, Eric E.; Kubiak, Cllifford P.; Sathrum, Aaron J.; Smieja, Jonathan M. (2009). "Electrocatalytic and homogeneous approaches to conversion of CO2 to liquid fuels". Chem. Soc. Rev. 38: 89–99. doi:10.1039/b804323j.
7. Halmann, M. (1978). "Photoelectrochemical reduction of aqueous carbon dioxide on p-type gallium phosphide in liquid junction solar cells". Nature. 275 (5676): 115–116. doi:10.1038/275115a0.
8. Aurian-Blajeni, B.; Halmann, M.; Manassen, J. (1983). "Electrochemical measurement on the photoelectrochemical reduction of aqueous carbon dioxide on p-Gallium phosphide and p-Gallium arsenide semiconductor electrodes". Solar Energy Materials. 8 (4): 425–440. doi:10.1016/0165-1633(83)90007-2.
9. Yoneyama, Hiroshi; Sugimura, Kenji; Kuwabata, Susumu (1988). "Effects of electrolytes on the photoelectrochemical reduction of carbon dioxide at illuminated p-type cadmium telluride and p-type indium phosphide electrodes in aqueous solutions". J. Electroanal. Chem. Interfacial Electrochem. 249 (1-2): 143–153. doi:10.1016/0022-0728(88)80355-3.
10. Junfu, Liu; Baozhu, ChunYu (1992). "Photoelectrochemical reduction of carbon dioxide on a p+/p-Si photocathode in aqueous electrolyte". J. Electroanal. Chem. 324: 191–200. doi:10.1016/0022-0728(92)80045-6.
11. Hori, Yoshio; Wakebe, Hidetoshi; Tsukamoto, Toshio; Koga, Osamu (1994). "Electrocatalytic process of CO selectivity in electrochemical reduction of CO2 at metal electrodes in aqueous media". Electrochim. Acta. 39 (11-12): 1833–1839. doi:10.1016/0013-4686(94)85172-7.
12. Kumar, Bhupendra; Smieja, Jonathan M.; Kubiak, Clifford P. (2010). "Photoreduction of CO2 on p-type Silicon Using Re(bipy-But)(CO)3Cl: Photovoltages Exceeding 600 mV for the Selective Reduction of CO2 to CO". J. Phys. Chem. C. 114 (33): 14220-14223. doi:10.1021/jp105171b.
13. Walter, Michael G.; Warren, Emily L.; McKone, James R.; Boettcher, Shannon W.; Mi, Qixi; Santori, Elizabeth A.; Lewis, Nathan S. (2010). "Solar Water Splitting Cells". Chem. Rev. 110 (11): 6446-6473. doi:10.1021/cr1002326.
14. Kaneco, Satoshi; Katsumata, Hideyuki; Suzuki, Tohru; Ohta, Kiyohisa (2006). "Photoelectrochemical reduction of carbon dioxide at p-type gallium arsenide and p-type indium phosphide electrodes in methanol". Chemical Engineering Journal. 116 (3): 227–231. doi:10.1016/j.cej.2005.12.014.
15. Taniguchi, I.; Aurian-Blajeni, B.; Bockris, J. O'M. (1984). "The reduction of carbon dioxide at illuminated p-type semiconductor electrodes in nonaqueous media". Electrochim. Acta. 29 (7): 923–932. doi:10.1016/0013-4686(84)87137-6.
16. Lovley, Derek (May 26, 2010). "Microbial Electrosynthesis: Feeding Microbes Electricity To Convert Carbon Dioxide and Water to Multicarbon Extracellular Organic Compounds". mBio. 1 (2): e00103–10. doi:10.1128/mBio.00103-10. PMID 20714445.

17. Reece, Steven Y.; Hamel, Jonathan A.; Sung, Kimberly; Jarvi, Thomas D.; Esswein, Arthur J.; Pijpers, Joep J. H.; Nocera, Daniel G. (November 4, 2011). "Wireless Solar Water Splitting Using Silicon-Based Semiconductors and Earth-Abundant Catalysts". Science. 334 (6056): 645–648. doi:10.1126/science.1209816. PMID 21960528.
18. "ELECTROFUELS: Microorganisms for Liquid Transportation Fuel". ARPA-E. Retrieved July 23, 2013.
19. "Novel Biological Conversion of Hydrogen and Carbon Dioxide Directly into Free Fatty Acids". ARPA-E. Archived from the original on October 10, 2013. Retrieved July 23, 2013.
20. "Electrofuels Via Direct Electron Transfer from Electrodes to Microbes". ARPA-E. Archived from the original on October 10, 2013. Retrieved July 23, 2013.
21. "SBE's Conference on Electrofuels Research". American Institute of Chemical Engineers. Retrieved July 23, 2013.
22. Biello, David (March 20, 2014). "Fracking Hammers Clean Energy Research". Scientific American. Retrieved April 14, 2014. The cheap natural gas freed from shale by horizontal drilling and hydraulic fracturing (or fracking) has helped kill off bleeding-edge programs like Electrofuels, a bid to use microbes to turn cheap electricity into liquid fuels, and ushered in programs like REMOTE, a bid to use microbes to turn cheap natural gas into liquid fuels.
23. Continuous Electrodeionization (CEDI/EDI) Ionpure CEDI Products
24. CEDI University web site
25. Electrodeionization (EDI), The Dow Chemical Company
26. Electrodeionization Systems, Electrodeionization Systems
27. Advanced Electrodeionization Technology for Product Desalting, Argonne National Laboratory
28. , Resistivity / Conductivity Measurement of Purified Water, Lab Manager Magazine
29. Gabriele Centi, Siglinda Perathoner "Opportunities and prospects in the chemical recycling of carbon dioxide to fuels" Catalysis Today 148 (2009) 191–205. doi:10.1016/j.cattod.2009.07.075
30. J. Qiao, et al., A review of catalysts for the electroreduction of carbon dioxide to produce low-carbon fuels, Chem.Soc.Rev., 2014, 43 , 631-675.
31. Appel, A. M. et al. "Frontiers, Opportunities, and Challenges in Biochemical and Chemical Catalysis of CO2 Fixation", Chem. Rev. 2013, vol. 113, 6621-6658. doi:10.1021/cr300463y
32. J. C. Fontecilla-Camps, P. Amara, C. Cavazza, Y. Nicolet and A. Volbeda, "Structure-function relationships of anaerobic gas-processing metalloenzymes", Nature 2009, volume 460, p. 814-822.doi:10.1038/nature08299
33. Susan Topham, "Carbon Dioxide" in Ullmann's Encyclopedia of Industrial Chemistry, 2005, Wiley-VCH, Weinheim. doi:10.1002/14356007.a05_165

34. S. Lee et al., ChemSusChem 9 (2016) 333-344, Electrode Build-Up of Reducible Metal Composites toward Achievable Electrochemical Conversion of Carbon Dioxidedoi:10.1002/cssc.201501112
35. S. Lee et al., Sustainable production of formic acid by electrolytic reduction of gaseous carbon dioxide, J. Mater. Chem. A, 2015,3, 3029-3034 doi:10.1039/C4TA03893B
36. D.T. Whipple et al., Prospects of CO2 Utilization via Direct Heterogeneous Electrochemical Reduction J. Phys. Chem. Lett., 2010, 1 (24), pp 3451–3458 doi:10.1021/jz1012627
37. R.L. Machundaa et al., Electrocatalytic reduction of CO2 gas at Sn based gas diffusion electrode, Current Applied Physics 11 (2011) 986. doi:10.1016/j.cap.2011.01.003
38. Hori, Y. (2008). "Electrochemical CO2 Reduction on Metal Electrodes". Modern Aspects of Electrochemistry. Modern Aspects of Electrochemistry. 42. pp. 89–80. doi:10.1007/978-0-387-49489-0_3. ISBN 978-0-387-49488-3.
39. Eric E. Benson, Clifford P. Kubiak,* Aaron J. Sathrum and Jonathan M. Smieja "Electrocatalytic and homogeneous approaches to conversion of CO2 to liquid fuels" Chem. Soc. Rev., 2009, vol. 38, pp. 89–99.doi:10.1039/b804323j
40. Halmann and Steinberg, "Greenhouse Gas Carbon Dioxide Mitigation," Lewis Publishers, 1999. ISBN 1-56670-284-4
41. Microbial Electrosynthesis: Feeding Microbes Electricity To Convert Carbon Dioxide and Water to Multicarbon Extracellular Organic Compounds
42. Microbial Electrolysis Cell - Turning Bacteria Into Hydrogen Machines
43. Microbial electrosynthesis — revisiting the electrical route for microbial production : Abstract : Nature Reviews Microbiology
44. Badwal, SPS (2014). "Emerging electrochemical energy conversion and storage technologies". Frontiers in Chemistry. 79. doi:10.3389/fchem.2014.00079. PMC 4174133Freely accessible. PMID 25309898.
45. Min, B.; Cheng, S.; Logan, B. E. (2005). "Electricity generation using membrane and salt bridge microbial fuel cells". Water Research. 39 (9): 1675–86. doi:10.1016/j.watres.2005.02.002.
46. "MFC Pilot plant at the Fosters Brewery". Retrieved 2013-03-09.
47. Potter, M.C. Potter (1911). "Electrical effects accompanying the decomposition of organic compounds". Royal Society (Formerly Proceedings of the Royal Society) B. 84: 260–276. doi:10.1098/rspb.1911.0073.
48. Cohen, B (1931). "The Bacterial Culture as an Electrical Half-Cell". Journal of Bacteriology. 21: 18–19.
49. DelDuca, M. G., Friscoe, J. M. and Zurilla, R. W. (1963). Developments in Industrial Microbiology. American Institute of Biological Sciences, 4, pp81–84.
50. Karube, I.; Matasunga, T.; Suzuki, S.; Tsuru, S. (1976). "Continuous hydrogen production by immobilized whole cels of Clostridium butyricum". Biocheimica et Biophysica Acta. 24 (2): 338–343.

51. Karube, Isao; Matsunaga, Tadashi; Tsuru, Shinya; Suzuki, Shuichi (November 1977). "Biochemical cells utilizing immobilized cells of Clostridium butyricum". Biotechnology and Bioengineering. 19 (11): 1727–1733. doi:10.1002/bit.260191112.

52. "Brewing a sustainable energy solution". The University of Queensland Australia. Retrieved 26 August 2014.

53. Allen, R.M.; Bennetto, H.P. (1993). "Microbial fuel cells: Electricity production from carbohydrates". Applied Biochemistry and Biotechnology. 39-40: 27–40. doi:10.1007/bf02918975.

54. Pant, D.; Van Bogaert, G.; Diels, L.; Vanbroekhoven, K. (2010). "A review of the substrates used in microbial fuel cells (MFCs) for sustainable energy production". Bioresource Technology. 101 (6): 1533–43. doi:10.1016/j.biortech.2009.10.017. PMID 19892549.

55. Lu, Z.; Chang, D.; Ma, J.; Huang, G.; Cai, L.; Zhang, L. (2015). "Behavior of metal ions in bioelectrochemical systems: A review". Journal of Power Sources. 275: 243–260. doi:10.1016/j.jpowsour.2014.10.168.

56. Oh, S.; Logan, B. E. (2005). "Hydrogen and electricity production from a food processing wastewater using fermentation and microbial fuel cell technologies". Water Research. 39 (19): 4673–4682. doi:10.1016/j.watres.2005.09.019.

57. Subhas C Mukhopadhyay; Joe-Air Jiang (2013). Wireless Sensor Networks and Ecological Monitoring. Springer link. pp. 151–178. ISBN 978-3-642-36365-8.

58. Chua SL, Wang VB, Cai Z, Sivakumar K, Kjelleberg S, Cao B, Loo SC, Yang L (2014). "A stable synergistic microbial consortium for simultaneous azo dye removal and bioelectricity generation". Bioresource technology. 155: 71–6. doi:10.1016/j.biortech.2013.12.078. PMID 24434696.

59. Chua SL, Wang VB, Cao B, Loo SC, Yang L (2013). "A stable synergistic microbial consortium for simultaneous azo dye removal and bioelectricity generation". PLOS ONE. 8: e63129. doi:10.1371/journal.pone.0063129. PMC 3659106Freely accessible. PMID 23700414.

60. Venkata Mohan, S.; Mohanakrishna, G.; Srikanth, S.; Sarma, P. N. (2008). "Harnessing of bioelectricity in microbial fuel cell (MFC) employing aerated cathode through anaerobic treatment of chemical wastewater using selectively enriched hydrogen producing mixed consortia" (PDF). Fuel. 87 (12): 2667–2676. doi:10.1016/j.fuel.2008.03.002.

61. Venkata Mohan, S.; Mohanakrishna, G.; Reddy, B. P.; Saravanan, R.; Sarma, P. N. (2008). "Bioelectricity generation from chemical wastewater treatment in mediatorless (anode) microbial fuel cell (MFC) using selectively enriched hydrogen producing mixed culture under acidophilic microenvironment" (PDF). Biochemical Engineering Journal. 39: 121–130. doi:10.1016/j.bej.2007.08.023.

62. Venkata Mohan, S.; Veer Raghuvulu, S.; Srikanth, S.; Sarma, P.N. (2007). "Bioelectricity production by meditorless microbial fuel cell (MFC) under acidophilic condition using wastewater as substrate: influence of substrate loading rate". Current Sci. 92 (12): 1720–1726.

63. Venkata Mohan, S.; Saravanan, R.; Raghavulu, S. V.; Mohanakrishna, G.; Sarma, P. N. (2008). "Bioelectricity production from wastewater treatment in dual-chamber microbial fuel cell (MFC) using selectively enriched mixed microflora: Effect of catholyte" (PDF). Bioresource Technology. 99 (3): 596–603. doi:10.1016/j.biortech.2006.12.026. PMID 17321135.
64. Venkata Mohan, S.; Veer Raghavulu, S.; Sarma, P. N. (2008). "Biochemical evaluation of bioelectricity production process from anaerobic wastewater treatment in a single-chamber microbial fuel cell (MFC) employing glass wool membrane" (PDF). Biosensors and Bioelectronics. 23 (9): 1326–1332. doi:10.1016/j.bios.2007.11.016. PMID 18248978.
65. Venkata Mohan, S.; Veer Raghavulu, S.; Sarma, P. N. (2008). "Influence of anodic biofilm growth on bioelectricity production in single-chamber mediatorless microbial fuel cell using mixed anaerobic consortia" (PDF). Biosensors and Bioelectronics. 24 (1): 41–47. doi:10.1016/j.bios.2008.03.010. PMID 18440217.
66. Choi Y., Jung S. and Kim S. (2000) Development of Microbial Fuel Cells Using Proteus Vulgaris Bulletin of the Korean Chemical Society, 21 (1), pp44–48
67. Yue & Lowther, 1986
68. Chen, T.; Barton, S.C.; Binyamin, G.; Gao, Z.; Zhang, Y.; Kim, H.-H.; Heller, A. (Sep 2001). "A miniature biofuel cell.". J Am Chem Soc. 123 (35): 8630–1. doi:10.1021/ja0163164. PMID 11525685.
69. Bullen RA, Arnot TC, Lakeman JB, Walsh FC (2006). "Biofuel cells and their development.". Biosensors & Bioelectronics. 21 (11): 2015–45. doi:10.1016/j.bios.2006.01.030. PMID 16569499.
70. Eos magazine, Waterstof uit het riool, June 2008
71. MudWatt. "MudWatt Science Kit". MudWatt.
72. Kim, BH.; Chang, IS.; Gil, GC.; Park, HS.; Kim, HJ. (April 2003). "Novel BOD (biological oxygen demand) sensor using mediator-less microbial fuel cell.". Biotechnology Letters. 25 (7): 541–545. doi:10.1023/A:1022891231369. PMID 12882142.
73. Chang, I. S.; Moon, H.; Jang, J. K.; Kim, B. H. (2005). "Improvement of a microbial fuel cell performance as a BOD sensor using respiratory inhibitors" (PDF). Biosensors and Bioelectronics. 20 (9): 1856–1859. doi:10.1016/j.bios.2004.06.003. PMID 15681205.
74. Gong, Y., Radachowsky, S. E., Wolf, M., Nielsen, M. E., Girguis, P. R., & Reimers, C. E. (2011). "Benthic Microbial Fuel Cell as Direct Power Source for an Acoustic Modem and Seawater Oxygen/Temperature Sensor System". Environmental Science and Technology. 45 (11): 5047–53. doi:10.1021/es104383q.
75. Biffinger, J.C., Little, B., Pietron, J., Ray, R., Ringeisen, B.R. (2008). "Aerobic Miniature Microbial Fuel Cells". NRL Review: 141–42.
76. Heijne, Annemiek Ter; Liu, Fei; Weijden, Renata van der; Weijma, Jan; Buisman, Cees J.N.; Hamelers, Hubertus V.M. (2010). "Copper Recovery Combined with Electricity Production in a Microbial Fuel Cell". Environmental Science &

Technology. 44 (11): 4376–4381. doi:10.1021/es100526g.

77. Heidrick, E S; J. Dolfing; K. Scott; S. R. Edwards; C. Jones; T. P. Curtis (2013). "Production of hydrogen from domestic wastewater in a pilot-scale microbial electrolysis cell". Applied Microbiology and Biotechnology. 97 (15): 6979–6989. doi:10.1007/s00253-012-4456-7.

78. Zhang, Fei, He, Zhen, Ge, Zheng (2013). "Using Microbial Fuel Cells to Treat Raw Sludge and Primary Effluent for Bioelectricity Generation". Department of Civil Engineering and Mechanics; University of Wisconsin - Milwaukee.

79. Delaney, G. M.; Bennetto, H. P.; Mason, J. R.; Roller, S. D.; Stirling, J. L.; Thurston, C. F. (2008). "Electron-transfer coupling in microbial fuel cells. 2. Performance of fuel cells containing selected microorganism-mediator-substrate combinations". Journal of Chemical Technology and Biotechnology. Biotechnology. 34: 13–27. doi:10.1002/jctb.280340104.

80. Lithgow, A.M., Romero, L., Sanchez, I.C., Souto, F.A., and Vega, C.A. (1986). Interception of electron-transport chain in bacteria with hydrophilic redox mediators. J. Chem. Research, (S):178–179.

81. Kim, B.H.; Kim, H.J.; Hyun, M.S.; Park, D.H. (1999a). "Direct electrode reaction of Fe (III) reducing bacterium, Shewanella putrefacience" (PDF). J Microbiol. Biotechnol. 9: 127–131.

82. Pham, C. A.; Jung, S. J.; Phung, N. T.; Lee, J.; Chang, I. S.; Kim, B. H.; Yi, H.; Chun, J. (2003). "A novel electrochemically active and Fe(III)-reducing bacterium phylogenetically related to Aeromonas hydrophila, isolated from a microbial fuel cell". FEMS Microbiology Letters. 223 (1): 129–134. doi:10.1016/S0378-1097(03)00354-9. PMID 12799011.

83. Mediator-less microbial fuel cell schematic + explanation Archived March 10, 2011, at the Wayback Machine.

84. "Environmental Technology". Wageningen UR.

85. Strik, David; H. V. M. Hamelers; Jan F. H. Snel; Cees J. N. Buisman (July 2008). "Green electricity production with living plants and bacteria in a fuel cell". International Journal of Energy Research. 32 (9): 870–876. doi:10.1002/er.1397.

86. "Advanced Water Management Centre". Advanced Water Management Centre.

87. "DailyTech - Microbial Hydrogen Production Threatens Extinction for the Ethanol Dinosaur".

88. Nevin Kelly P.; Woodard Trevor L.; Franks Ashley E.; et al. (May–June 2010). "Microbial Electrosynthesis: Feeding Microbes Electricity To Convert Carbon Dioxide and Water to Multicarbon Extracellular Organic Compounds". mBio. 1 (2): e00103–10. doi:10.1128/mBio.00103-10. PMID 20714445.

89. Xu, Bojun; Ge, Zheng; He, Zhen (2015). "Sediment microbial fuel cells for wastewater treatment: challenges and opportunities,". Environ. Sci.: Water Res. Technol. The Royal Society of Chemistry,. 1: –,. doi:10.1039/C5EW00020C. Retrieved April 2015. Check date values in: |access-date= (help)

90. Clark, Helen (March 2, 2015). "Cleaning up wastewater from oil and gas operations using a microbe-powered battery". Gizmag. Check date values in: |access-date= (help);

91. Elizabeth, Elmy (2012). "GENERATING ELECTRICITY BY "NATURE'S WAY"". SALT 'B' online magazine. 1.

92. Strik, David; Ruud A. Timmers; Marjolein Helder; Kirsten J.J. Steinbusch; Hubertus V.M. Hamelers; Cees J.N. Buisman (2011). "Microbial solar cells: applying photosynthetic and electrochemically active organisms". Trends in Biotechnology. 29: 41–49. doi:10.1016/j.tibtech.2010.10.001.

93. Bombelli, Paolo; Bradley, Robert W.; Scott, Amanda M.; Philips, Alexander J.; McCormick, Alistair J.; Cruz, Sonia M.; Anderson, Alexander; Yunus, Kamran; Bendall, Derek S.; Cameron, Petra J.; Davies, Julia M.; Smith, Alison G.; Howe, Christopher J.; Fisher, Adrian C. (2011). "Quantitative analysis of the factors limiting solar power transduction by Synechocystis sp. PCC 6803 in biological photovoltaic devices". Energy & Environmental Science. 4 (11): 4690–4698. doi:10.1039/c1ee02531g.

94. "Miniature Microbial Fuel Cells". Technology Transfer Office. Retrieved 30 November 2014.

95. Biffinger, Justin C.; Ray, Ricky; Little, Brenda; Ringeisen, Bradley R. (2007). "Diversifying Biological Fuel Cell Design by Use of Nanoporous Filters". Environmental Science and Technology. 41 (4): 1444–49. doi:10.1021/es061634u.

96. Shabeeba, Anthru (5 Jan 2016). "Seminar 2". Slide Share.

97. Pasternak, Grzegorz; Greenman, John; Ieropoulos, Ioannis (2016-01-01). "Comprehensive Study on Ceramic Membranes for Low-Cost Microbial Fuel Cells". ChemSusChem. 9 (1): 88–96. doi:10.1002/cssc.201501320. ISSN 1864-564X.

98. Behera, Manaswini; Jana, Partha S.; Ghangrekar, M. M. (2010-02-01). "Performance evaluation of low cost microbial fuel cell fabricated using earthen pot with biotic and abiotic cathode". Bioresource Technology. 101 (4): 1183–1189. doi:10.1016/j.biortech.2009.07.089.

99. Winfield, Jonathan; Greenman, John; Huson, David; Ieropoulos, Ioannis (2013-06-01). "Comparing terracotta and earthenware for multiple functionalities in microbial fuel cells". Bioprocess and Biosystems Engineering. 36 (12): 1913–1921. doi:10.1007/s00449-013-0967-6. ISSN 1615-7591.

100. Bennetto, H. P. (1990). "Electricity Generation by Micro-organisms" (PDF). Biotechnology Education. 1 (4): 163–168.

101. Cheng, Ky; Ho, G; Cord-Ruwisch, R. "Affinity of microbial fuel cell biofilm for the anodic potential.". Environmental Science & Technology. 42 (10): 3828–34. doi:10.1021/es8003969. ISSN 0013-936X.

102. Bennetto, HP.; Stirling, JL.; Tanaka, K.; Vega, CA. (Feb 1983). "Anodic reactions in microbial fuel cells.". Biotechnology and Bioengineering. 25 (2): 559–68. doi:10.1002/bit.260250219. PMID 18548670.

CHAPTER 10

Bioelectrochemistry

Chapter Contents

INTRODUCTION

Bioelectrochemistry is a branch of electrochemistry and biophysical chemistry concerned with electrophysiological topics like cell electron-proton transport, cell membrane potentials and electrode reactions of redox enzymes.

HISTORY

The beginnings of bioelectrochemistry, as well as those of electrochemistry, are closely related to physiology through the works of Luigi Galvani and then Alessandro Volta. The first modern work in this field is considered that of the German physiologist Julius Bernstein (1902) concerning the source of biopotentials due to different ion concentration through the cell's membrane. The domain of bioelectrochemistry has grown considerably over the past century, maintaining the close connections to various medical and biological and engineering disciplines like electrophysiology and biomedical engineering. The achievements in this field have been awarded several Nobel prizes for Physiology or Medicine. Among prominent electrochemists who have contributed to this field one could mention John Bockris.

BIOELECTROMAGNETICS

Bioelectromagnetics, also known as bioelectromagnetism, is the study of the interaction between electromagnetic fields and biological entities. Areas of study include electrical or electromagnetic fields produced by living cells, tissues or organisms, including bioluminescent bacteria; for example, the cell membrane potential and the electric currents that flow in nerves and muscles, as a result of action potentials. Others include animal navigation utilizing the geomagnetic field; potential effects of man-made sources of electromagnetic fields like mobile phones; and developing new therapies to treat various conditions. The term can also refer to the ability of living cells, tissues, and organisms to produce electrical fields and the response of cells to electromagnetic fields.

Biological Phenomena

Short-lived electrical events called action potentials occur in several types of animal cells which are called excitable cells, a category of cell include neurons, muscle cells, and endocrine cells, as well as in some plant cells. These action potentials are used to facilitate inter-cellular communication and activate intracellular processes. The physiological phenomena of action potentials are possible because voltage-gated ion channels allow the resting potential caused by electrochemical gradient on either side of a cell membrane to resolve.

Bioelectromagnetism is studied primarily through the techniques of electrophysiology. In the late eighteenth century, the Italian physician and physicist Luigi Galvani first recorded the phenomenon while dissecting a frog at a table where he had been conducting experiments with static electricity. Galvani coined the term *animal electricity* to describe the phenomenon, while contemporaries labeled it galvanism. Galvani and contemporaries regarded muscle activation as resulting from an electrical fluid or substance in the nerves.

Some usually aquatic animals, such as sharks, have acute bioelectric sensors providing a sense known as electroreception, while migratory birds navigate in part by orienteering with respect to the Earth's magnetic field. In an extreme application of electromagnetism the electric eel is able to generate a large electric field outside its body used for hunting and self-defense through a dedicated electric organ.

Thermal Effects

Most of the molecules in the human body interact weakly with electromagnetic fields in the radiofrequency or extremely low frequency bands. One such interaction is absorption of energy from the fields, which can cause tissue to heat up; more intense fields will produce greater heating. This can lead to biological effects ranging from muscle relaxation (as produced by a diathermy device) to burns. Many nations and regulatory bodies like the International Commission on Non-Ionizing Radiation Protection have established safety guidelines to limit EMF exposure to a non-thermal level. This can be defined as either heating only to the point where the excess heat can be dissipated, or as a fixed increase in temperature not detectable with current instruments like 0.1 °C. However, biological effects have been shown to be present for these non-thermal exposures; Various mechanisms have been proposed to explain these, and there may be several mechanisms underlying the differing phenomena observed. Biological effects of weak electromagnetic fields are the subject of study in magnetobiology.

Behavioral Effects

Many behavioral effects at different intensities have been reported from exposure to magnetic fields, particularly with pulsed magnetic fields. The specific pulseform used appears to be an important factor for the behavioural effect seen; for example, a pulsed magnetic field originally designed for spectroscopic MRI, referred to as Low Field Magnetic Stimulation, was found to temporarily improve patient-reported mood in bipolar patients, while another MRI pulse had no effect. A whole-body exposure to a pulsed magnetic field was found to alter standing balance and pain perception in other studies.

TMS and Related Effects

A strong changing magnetic field can induce electrical currents in conductive tissue such as the brain. Since the magnetic field penetrates tissue, it can be generated outside of the head to induce currents within, causing transcranial magnetic stimulation (TMS). These currents depolarize neurons in a selected part of the brain, leading to changes in the patterns of neural activity. In repeated pulse TMS therapy or rTMS, the presence of incompatible EEG electrodes can result in electrode heating and, in severe cases, skin burns. A number of scientists and clinicians are attempting to use TMS to replace electroconvulsive therapy (ECT) to treat disorders such as severe depression. Instead of one strong electric shock through the head as in ECT, a large number of relatively weak pulses are delivered in TMS therapy, typically at the rate of about 10 pulses per second. If very strong pulses at a rapid rate are delivered to the brain, the induced currents can cause convulsions much like in the original electroconvulsive therapy. Sometimes, this is done deliberately in order to treat depression, such as in ECT.

Health Effects

While health effects from extremely low frequency (ELF) electric and magnetic fields (0 to 300 Hz) generated by power lines, and radio/microwave frequencies (RF) (10 MHz - 300 GHz) emitted by radio antennas and wireless networks have been well studied, the intermediate range (IR) used increasingly in modern telecommunications (300 Hz to 10 MHz) has been studied far less. Direct effects of electromagnetism on human health have been difficult to prove, and documented life-threatening interferences from electromagnetic fields are limited to medical devices such as pacemakers and other electronic implants. However, a number of studies have been conducted with artificial magnetic fields and electric fields to investigate their effects on cell metabolism, apoptosis, and tumor growth.

Electromagnetic radiation in the intermediate frequency range has found a place in modern medical practice for the treatment of bone healing and for nerve stimulation and regeneration. It is also approved as cancer therapy in form of Tumor Treating Fields, using alternating electric fields in the frequency range of 100–300 kHz. Since some of these methods involve magnetic fields that invoke electric currents in biological tissues and others only involve electric fields, they are strictly speaking electrotherapies albeit their application modi with modern electronic equipment have placed them in the category of bioelectromagnetic interactions.

ELECTROCHEMILUMINESCENCE

Electrochemiluminescence or electrogenerated chemiluminescence (ECL) is a kind of luminescence produced during electrochemical reactions in solutions. In electrogenerated chemiluminescence, electrochemically generated intermediates undergo a highly exergonic reaction to produce an electronically excited state that then emits light upon relaxation to a lower-level state. This wavelength of the emitted photon of light corresponds to the energy gap between these two states. ECL excitation can be caused by energetic electron transfer (redox) reactions of electrogenerated species. Such luminescence excitation is a form of chemiluminescence where one/all reactants are produced electrochemically on the electrodes.

ECL is usually observed during application of potential (several volts) to electrodes of electrochemical cell that contains solution of luminescent species (polycyclic aromatic hydrocarbons, metal complexes, Quantum Dots or Nanoparticles) in aprotic organic solvent (ECL composition). In organic solvents both oxidized and reduced forms of luminescent species can be produced at different electrodes simultaneously or at a single one by sweeping its potential between oxidation and reduction. The excitation energy is obtained from recombination of oxidized and reduced species.

In aqueous medium, which is mostly used for analytical applications, simultaneous oxidation and reduction of luminescent species is difficult to achieve due to electrochemical splitting of water itself so the ECL reaction with the coreactants is used. In the later case luminescent species are oxidized at the electrode together with the coreactant which gives a strong reducing agent after some chemical transformations (the oxidative reduction mechanism).

Applications

ECL proved to be very useful in analytical applications as a highly sensitive and selective method. It combines analytical advantages of chemiluminescent analysis (absence of background optical signal) with ease of reaction control by applying electrode potential. As an analytical technique it presents outstanding advantages over other common analytical methods due to its versatility, simplified optical setup compared with photoluminescence (PL), and good temporal and spatial control compared with chemiluminescence (CL). Enhanced selectivity of ECL analysis is reached by variation of electrode potential thus controlling species that are oxidized/reduced at the electrode and take part in ECL reaction.

It generally uses Ruthenium complexes, especially $[Ru\ (Bpy)_3]^{2+}$ (which releases a photon at ~620 nm) regenerating with TPA (Tripropylamine) in liquid phase or liquid–solid interface. It can be used as monolayer immobilized on an electrode surface (made e.g. of nafion, or special thin films made by Langmuir–Blogett technique or self-assembly technique) or as a coreactant or more commonly as a tag and used in HPLC, Ru tagged antibody based immunoassays, Ru Tagged DNA probes for PCR etc., NADH or H_2O_2 generation based biosensors, oxalate and organic amine detection and many other applications and can be detected from picomolar sensitivity to dynamic range of more than six orders of magnitude. Photon detection is done with photomultiplier tubes (PMT) or silicon photodiode or gold coated fiber-optic sensors.The importance of ECL techniques detection for bio-related applications has been well established. ECL is heavily used commercially for many clinical lab applications.

BIPOLAR ELECTROCHEMISTRY

Bipolar electrochemistry is a phenomenon in electrochemistry based on the polarization of conducting objects in electric fields. Indeed, this polarization generates a potential difference between the two extremities of the substrate that is equal to the electric field value multiplied by the size of the object. If this potential difference is important enough, then redox reactions can be generated at the extremities of the object, oxidations will occur at one extremity coupled simultaneously to reductions at the other extremity. In a simple experimental setup consisting of a platinum wire in a weighing boat containing a pH indicator solution, a 30 V voltage across two electrodes will cause water reduction at one end of the wire (the cathode) and a pH increase (OH^- formation) and water oxidation at the anodic end and a pH decrease. The poles of the bipolar electrode also align themselves with the applied electric field.

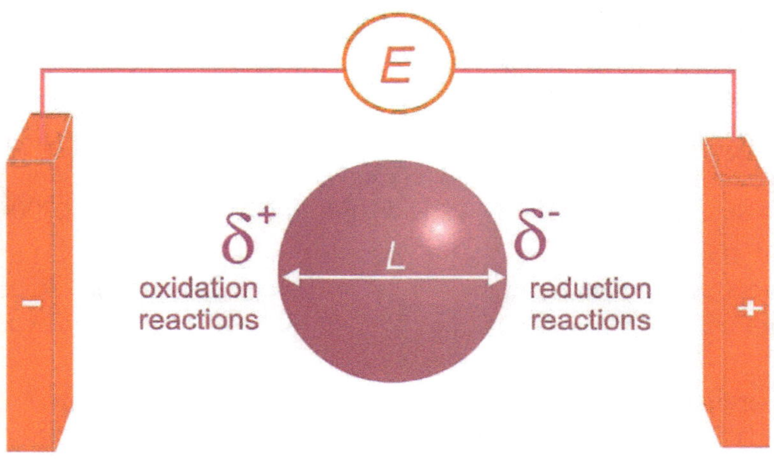

Bipolar electrochemistry concept.

Utilizations

The phenomenon of bipolar electrochemistry is known for several decades and is used in industry in some electrolytic reactors. The interest of the scientific community for this concept seems to increase a lot since Martin Fleischmann and co-workers demonstrated that water splitting was possible using micrometer-sized bipolar electrodes. Recently, several applications in such domains as synthesis of dissymmetrical micro- and nano-structures analytical chemistry material science, microelectronics and microobject propulsion have been developed.

REFERENCES

1. Electrochemistry Encyclopedia
2. "Bioelectrochemical Society".
3. Malmivuo, Jaakko; Robert Plonsey (1994). Bioelectromagnetism : principles and applications of bioelectric and biomagnetic fields. New York: Oxford University Press. ISBN 978-0195058239.
4. Myers, Richard (2003). The basics of chemistry. Westport, Conn.: Greenwood Press. pp. 172–174. ISBN 9780313316647.
5. "Hazards of the MR Environment". Martinos Center for Biomedical Imaging. Retrieved 19 March 2013.
6. Binhi, 2002
7. Rohan, M.; Parow, A; Stoll, AL; et al. (2004). "Low-Field Magnetic Stimulation in Bipolar Depression Using an MRI-Based Stimulator". American Journal of Psychiatry. 161 (1): 93–98. doi:10.1176/appi.ajp.161.1.93. PMID 14702256.
8. Thomas, A.W; White, K.P; Drost, D.J; Cook, C.M; Prato, F.S (2001). "A comparison of rheumatoid arthritis and fibromyalgia patients and healthy controls exposed to a pulsed (200 μT) magnetic field: effects on normal standing balance". Neuroscience Letters. 309 (1): 17–20. doi:10.1016/S0304-3940(01)02009-2.
9. Shupak, Naomi M; Prato, Frank S; Thomas, Alex W (2004). "Human exposure to a specific pulsed magnetic field: effects on thermal sensory and pain thresholds". Neuroscience Letters. 363 (2): 157–162. doi:10.1016/j.neulet.2004.03.069.
10. Roth, BJ; Pascual-Leone, A; Cohen, LG; Hallett, M (1992). "The heating of metal electrodes during rapid-rate magnetic stimulation: A possible safety hazard". Electroencephalography and Clinical Neurophysiology/Evoked Potentials Section. 85 (2): 116–23. doi:10.1016/0168-5597(92)90077-O. PMID 1373364.
11. Wassermann EM (1998). "Risk and safety of repetitive transcranial magnetic stimulation: report and suggested guidelines from the International Workshop on the Safety of Repetitive Transcranial Magnetic Stimulation, June 5–7, 1996" (pdf). Electroencephalography and clinical Neurophysiology. 108 (1): 1–16. doi:10.1016/S0168-5597(97)00096-8. PMID 9474057.
12. Rossi, S; Hallett, M; Rossini, PM; Pascual-Leone, A; Safety of TMS Consensus Group (2009). "Safety, ethical considerations, and application guidelines for the use of transcranial magnetic stimulation in clinical practice and research". Clinical Neurophysiology. 120 (12): 2008–2039. doi:10.1016/j.clinph.2009.08.016. PMC 3260536Freely accessible. PMID 19833552.
13. Electromagnetic fields & public health: Intermediate Frequencies (IF). Information sheet February 2005. World Health Organization. Retrieved Aug 2013.

14. Wartenberg, M., Wirtz, N., Grob, A., Niedermeier, W., Hescheler, J., Peters, S. C. and Sauer, H. (2008), "Direct current electrical fields induce apoptosis in oral mucosa cancer cells by NADPH oxidase-derived reactive oxygen species". Bioelectromagnetics, 29: 47–54. doi: 10.1002/bem.20361

15. Forster RJ, Bertoncello P, Keyes TE (2009). "Electrogenerated Chemiluminescence". Annual Review of Analytical Chemistry. 2: 359–85. Bibcode:2009ARAC....2..359F. doi:10.1146/annurev-anchem-060908-155305. PMID 20636067.

16. Electrogenerated Chemiluminescence, Edited by Allen J. Bard, Marcel Dekker, Inc., 2004

17. Valenti, G.; Bruno, C.; Rapino, S.; Fiorani, A.; Jackson, E.A.; Scott, L.T.; Paolucci, F.; Marcaccio, M. (October 2010). "Intense and Tunable Electrochemiluminescence of Corannulene". J. Phys. Chem. C. 114 (45): 19467–19472. doi:10.1021/jp107964r.

18. Rampazzo, E.; Bonacchi, S.; Genovese, D.; Juris, R.; Marcaccio, M.; Montalti, M.; Paolucci, F.; Sgarzi, M.; Valenti, G.; Zaccheroni, N.; Prodi, L. (August 2012). "Nanoparticles in metal complexes-based electrogenerated chemiluminescence for highly sensitive applications". Coordination Chemistry Reviews. 256 (15-16): 1664–1681. doi:10.1016/j.ccr.2012.03.021.

19. Fähnrich, K.A.; Pravda, M.; Guilbault, G. G. (May 2001). "Recent applications of electrogenerated chemiluminescence in chemical analysis" (PDF). Talanta. 54 (4): 531–559. doi:10.1016/S0039-9140(01)00312-5. PMID 18968276.[permanent dead link]

20. Miao, Wujian (2008). "Electrogenerated Chemiluminescence and Its Biorelated Applications". Chemical Reviews. 108 (7): 2506–2553. doi:10.1021/cr068083a.

21. Lee, Won-Yong (1997). "Tris (2,2′-bipyridyl)ruthenium(II) electrogenerated chemiluminescence in analytical science". Mikrochimica Acta. 127 (1-2): 19–39. doi:10.1007/BF01243160.

22. Wei, Hui; Wang, Erkang (2008-05-01). "Solid-state electrochemiluminescence of tris(2,2′-bipyridyl) ruthenium". TrAC Trends in Analytical Chemistry. 27 (5): 447–459. doi:10.1016/j.trac.2008.02.009.

23. Wei, Hui; Wang, Erkang (2011-03-01). "Electrochemiluminescence of tris(2,2′-bipyridyl)ruthenium and its applications in bioanalysis: a review". Luminescence. 26 (2): 77–85. doi:10.1002/bio.1279. ISSN 1522-7243.

24. G. Loget; A. Kuhn (2011). "Shaping and exploring the micro- and nanoworld using bipolar electrochemistry". Analytical and Bioanalytical Chemistry. 400 (6): 1691. doi:10.1007/s00216-011-4862-1.

25. F. Mavré; R. K. Anand; D. R. Laws; K.-F. Chow; B.-Y. Chang; J. A. Crooks; R. M. Crooks (2010). "Feature Bipolar Electrodes: A Useful Tool for Concentration, Separation, and Detection of Analytes in Microelectrochemical Systems". Anal. Chem. 82 (21): 8766. doi:10.1021/ac101262v.

26. Fosdick, S. E.; Knust, K. N.; Scida, K.; Crooks, R. M. (2013). "Bipolar Electrochemistry". Angew. Chem. Int. Ed. 52: 10438–10456. doi:10.1002/anie.201300947.

27. M. Fleischmann; J. Ghoroghchian; D. Rolison; S. Pons (1986). "Electrochemical behavior of dispersions of spherical ultramicroelectrodes". J. Phys. Chem. 90 (23): 6392. doi:10.1021/j100281a065.

28. G. Loget; V. Lapeyre; P. Garrigue; C. Warakulwit; J. Limtrakul; M.-H. Delville; A. Kuhn (2011). "Versatile Procedure for Synthesis of Janus-Type Carbon Tubes". Chem. Mater. 23 (10): 2595. doi:10.1021/cm2001573.

29. C. Warakulwit; T. Nguyen; J. Majimel; M.-H. Delville; V. Lapeyre; P. Garrigue; V.Ravaine; J. Limtrakul; A. Kuhn (2008). "Dissymmetric Carbon Nanotubes by Bipolar Electrochemistry". Nano Lett. 8 (2): 500. Bibcode:2008NanoL...8..500W. doi:10.1021/nl072652s.

30. K.-F. Chow; B.-Y. Chang; B. A. Zaccheo; F. Mavré; R. M. Crooks (2010). "A Sensing Platform Based on Electrodissolution of a Ag Bipolar Electrode". J. Am. Chem. Soc. 132 (27): 9228. doi:10.1021/ja103715u.

31. Hlushkou D, Perdue RK, Dhopeshwarkar R, Crooks RM, Tallarek U (2009). "Electric field gradient focusing in microchannels with embedded bipolar electrode". Lab Chip. 9 (13): 1903. doi:10.1039/b822404h.

32. Ulrich C, Andersson O, Nyholm L, Björefors F (2009). "Potential and Current Density Distributions at Electrodes Intended for Bipolar Patterning". Anal. Chem. 81: 453. doi:10.1021/ac801871c.

33. Ramakrishnan S, Shannon C (2010). "Display of Solid-State Materials Using Bipolar Electrochemistry". Langmuir. 26 (7): 4602. doi:10.1021/la100292u.

34. J. C. Bradley; H. M. Chen; J. Crawford; J. Eckert; K. Ernazarova; T. Kurzeja; M. Lin; M. McGee; W. Nadler; S. G. Stephens (1997). "Creating electrical contacts between metal particles using directed electrochemical growth". Nature. 389 (6648): 268. Bibcode:1997Natur.389..268B. doi:10.1038/38464.

35. G. Loget; A. Kuhn (2010). "Propulsion of Microobjects by Dynamic Bipolar Self-Regeneration". J. Am. Chem. Soc. 132 (45): 15918. doi:10.1021/ja107644x.

36. G. Loget; A. Kuhn (2011). "Electric field-induced chemical locomotion of conducting objects". Nature Communications. 2 (11): 535. Bibcode:2011NatCo...2E.535L. doi:10.1038/ncomms1550.

Index